The ancient Greeks disdained practical labor, and the technical arts as well; medieval Christians extolled labor, but looked askance at devices designed to reduce or eliminate it. Books as diverse as the Bible, which proceeds from the world's creation, and Aldous Huxley's *Brave New World*, set in 632 A.F. (After Ford), have warned of technology's dark potentials. Prometheus brought the gift of fire to humans, while Mary Shelley's Dr. Frankenstein created life and then abandoned it; both met bleak fates.

Barry Katz's *Technology and Culture: a Historical Romance* looks at these and many other tales of technology's role in the human quest for knowledge and power. Wielding the historical record like an archaeologist's pick, Katz excavates a field spanning five thousand years and stretching from the Tower of Babel to the Eiffel Tower. Readers will encounter skeptics and visionaries, metalsmiths and magicians, as Katz sifts through fragments of shattered texts, unearths bones of contention, and reconstructs the attitudes with which the art of practical invention has contended throughout history.

Technology and Culture:
A Historical Romance

The Portable Stanford Book Series

Published by the Stanford Alumni Association

THE PORTABLE STANFORD is a book series
sponsored by the Stanford Alumni Association.
The series is designed to bring the widest possible
sampling of Stanford's intellectual resources into the
homes of alumni. It includes books based on current
research as well as books that deal with philosophical
issues, which by their nature reflect to a greater degree
the personal views of their authors.

THE PORTABLE STANFORD BOOK SERIES
Stanford Alumni Association
Bowman Alumni House
Stanford, California 94305-4005

Library of Congress Catalog Card
Number: 90-71906
ISBN: 0-916318-45-1

10 9 8 7 6 5 4 3 2 1

Acknowledgments

■ ■ ■ ■ ■ ■

This book was conceived amid ongoing discussions with my wife, Deborah Trilling, a painter for whom the phrase "mechanical arts" is an oxymoron, and my friend Jim Adams, an engineer for whom it is redundant. I neither could nor would have written it without their yinning and yanging at me all these years. In fact, I will decline to add the usual disclaimer that frees one's acknowledgees from all responsibility: to the contrary, it is their fault.

The Stanford Program in Values, Technology, Science and Society houses a remarkable community of scholars who have provided an endless source of ideas, information, objections, and much more. I must ask the reader's indulgence if I thank each of the past and present colleagues with whom I have taught the course VTSS 1-2-3, "Technology and Culture," over the past ten years: Lois Becker, Sharona Ben-Tov, Iain Boal, Ronald Bracewell, Brigitte Comparini, Joseph Corn, Mark Edwards, Keith Gandal, Janet Gardiner, Edwin Good, David Horn, Peter Hirtle, Eric Hutchinson, Donald Jordan, Wilbur Knorr, Robert McGinn, Patricia Nabti, Curtis Runnells, Londa Schiebinger, Janet Schmidt, Paul Seaver, Sharon Traweek, Walter Vincenti, and Rosemary Wakeman.

Finally, special thanks go to Bruce Goldman and Amy Pilkington for their intelligent criticism, expert advice, and exemplary patience.

Series Editor and Manager: Bruce Goldman
Production Manager: Amy Pilkington
Cover Designer: Paul Carstensen

To Deborah: critic, collaborator, accomplice, and wife.

Table of Contents

3

Technology and World Culture

■ ■ ■ ■ ■ ■

Page 47

4

Head and Hand in the
Culture of the Renaissance

■ ■ ■ ■ ■ ■

Page 71

5

Technology and Revolution

■ ■ ■ ■ ■ ■

Page 99

6

The Prosthetic God

■ ■ ■ ■ ■ ■

Page 119

Conclusion:
A Historical Romance

■ ■ ■ ■ ■ ■

Page 143

Bibliography

■ ■ ■ ■ ■ ■

Page 145

Credits

■ ■ ■ ■ ■ ■

Page 153

About the Author

■ ■ ■ ■ ■ ■

Page 157

Introduction

■ ■ ■ ■ ■ ■

Technology and Culture

Technology is older than art, religion, or philosophy;

indeed, it is older than human beings themselves.

Many animals use objects taken directly from nature

to assist them, and it has even been argued that

tool-wielding apes may have so transformed

their environments as to call forth the enhanced mental capacities of *homo sapiens*. It would seem that the tools we have found, inherited, and fashioned are more than artifacts of our cultural history; they are factors in our biological evolution.

Given the centrality of technology to human existence, it is difficult to explain the tendency to segregate it from the works of art and intellect and even "disinterested" scientific speculation that is such a striking feature of history. Indeed, prior to our own century, when technology has come to be linked either to Salvation or to Apocalypse, it is the exceptions to this rule that command our attention. Francis Bacon remarked as early as 1620 that "no empire, no sect, no star" has exerted greater influence over human affairs than a handful of "mechanical discoveries"—printing, which diffused literacy throughout society; gunpowder, which sounded the death knell of the feudal aristocracy; and the magnetic needle, which enabled Europeans to set out across the open seas. In this respect, however (as in his indictment for bribe-taking), the Lord High Chancellor was prematurely modern, a lone visionary whose insight illuminates the path to the future more than the intellectual climate of the world in which he himself lived.

The relative neglect of technology by the leading writers and artists of past generations can be explained by two sorts of factors. First, and most obvious, is the oft-remarked aloofness of the intellectual elite from the manual trades, so bluntly exhibited by the aristocratic university president Plato: We can scarcely do without the services of engineers, he admits to his friend Callicles—"but would you let your daughter marry one?" Modern society has by no means shaken free of this ancient prejudice.

More decisive than this patrician disdain, however, is the fact that for the greater part of our history, technology never seemed powerful enough to exercise a pervasive influence. Only in the last two centuries has it commanded center stage: The design of our cities reflects not conscious decisions about how we ought to live, but arrangements made possible by the automobile, the skyscraper, the telephone; the innermost values of a generation are tied to a world that exists only within the frame of the television screen; whatever the wisdom or the folly of our governments, it is the very existence of the weapons they control that threatens us.

The emergence of technology as a force to be reckoned with on its own terms can be written as a historical romance. For most of antiquity, the practical activities of farmers, artisans, and surgeons lived apart from the tradition of abstract speculation about the good, the true, and the beautiful. They began to meet discreetly in the workshops of the famous Renaissance engineer-architects and to flirt audaciously in the laboratories of the early modern alchemists. Then, in the early 17th century, that old panderer Francis Bacon advised that "if you would control nature you must first understand her," and with the founding of the great national academies of that century theory and practice finally dared to court one another openly. Only in the maturity of the Industrial Revolution did they begin to cohabit, however, and it was not until modern times—starting first with the chemical and electrical industries of the late 19th century—that they became permanently wedded. The consummation of that marriage—manifest today in the infant biotechnology, microelectronics, and energy industries—has transformed technology from one human activity among many to a central force of dazzling, terrifying, and as yet inconceivable power.

The task of the following chapters will be to trace the diverse ways over time in which philosophers and scientists, painters and poets, theologians and political theorists have attempted to grasp the nature of technology and to assess the promises and dangers it seemed to hold. In some cases their positions will be clearly and even bluntly stated; more often, however, the changing attitudes of our ancestors toward technology lie buried in myth and poetry and philosophy, and await excavation. This is an intellectual archaeology, then, in search not of technologies but of texts. With luck, our expedition will provide a revealing insight into the past, but it may also yield results that tell us something about ourselves.

1

■ ■ ■ ■ ■ ■

From the Garden of Eden to the Tower of Babel: Technology and the Origins of Civilization

Around ten or twelve thousand years ago the

human race got itself committed to an

experiment called "civilization." After thousands

of generations of nomadic, hunter-gatherer

subsistence, groups of our ancestors entered upon

the bold venture of "living-together-in-cities." This experiment has occupied the merest fraction of the 2 $\frac{1}{2}$ million years we have spent on the planet, and it is, accordingly, far too early for us to judge whether it will work out. What is clear is that by virtue of the transformation of the globe it has enabled us to achieve, the experiment is irrevocable. The means of that transformation I will call "technology."

This simple definition encompasses much more than hardware—the tools, instruments, and machines with which we physically mold our surroundings. We could not even have begun the work of civilization without a comparable endowment of software: conceptions of nature and of how it works, religious values that tell us what constitutes the good life and how far we may go to attain it, the social controls necessary to mobilize the labor of individuals and groups, arrangements for communicating information across time and space, and much else. The traditional culture that assumes that a local deity inhabits every rock and field and forest is unlikely to develop heavy earth-moving equipment; Christians who believe, with St. Augustine, that "we are but sojourners in this world" may not care to invest much in the infrastructure of such a transient dwelling-place; rationalist followers of Descartes, by contrast, may see everything outside the thinking self as mere "extension," and available for measuring, surveying, excavating, enclosing, or developing. Needless to say, they too will develop the tools appropriate to their world view.

Historians of hardware search out their evidence in the usual places: deposits of fossilized bone tools, collections of early surgical instruments, abandoned factory buildings haunted by the ghosts of the workers who once toiled there. The *cultural* history of technology, however, which is as interested in what people thought as in what they actually did, has a wider but shorter field to cultivate: It plows through epic poetry and systematic theology, plans of ideal cities and political theories about how they should be governed, the warnings of poets and the promises of scientists. But it is in the nature of our evidence that we are rarely able to go back much before the invention of writing. Fortunately, this happens to be a good place to begin.

With the invention of writing, late in the fourth millennium B.C., human societies acquired the blessing—or the curse—of a cultural memory that extended more than two or three generations. Knowledge could become cumulative; that is to say, successive generations could build upon the intellectual accomplishments of their ancestors, and not simply receive and maintain them. Political and religious institutions requiring detailed record-keeping could flourish, and did so in the great river valleys of China, India, Egypt, and Mesopotamia. Economic life could be rationalized so as to allocate a precarious agricultural surplus among the options of consumption, investment (in next year's planting), and trade.

Writing, then, is part of the ongoing attempt by human beings to apprehend the universe and render it serviceable to their needs. No less than real estate contracts, "no trespassing" signs, or even the simple process of naming, the first writing was intended to isolate a portion of nature and bring it under human control. In traditional cultures, however, this implies a magical identification with the external world that is utterly foreign to our own scientific world view. Whereas moderns believe that the physical universe is an objective matrix, subject to its own laws and indifferent to our existence, the peoples of remote antiquity placed themselves within a continuum that included the heavens and the earth and everything in between. To capture the likeness of a thing in sound or image or symbol was to capture the thing itself.

The stages by which the technology of writing evolved have been reconstructed in exacting detail by specialists, and despite innumerable variations and exceptions, the basic sequence is pretty well established. The precursor of writing was the *pictogram*, which attempted simply to illustrate the sun, an ear of corn, a pregnant woman. As they became standardized, these figures became stylized, abbreviated, and diversified. Thus emerged the *ideogram*, which represented not a thing but a word—

"warmth," "food," or "fertility," to follow the above examples. A further development, which has proved vital in reconstructing the ideas and institutions of the ancient Near East, was Mesopotamian cuneiform: highly stylized, wedge-shaped impressions made in wet clay tablets which, when baked, are virtually indestructible.

In the heavily urbanized culture of the ancient Near East, after the middle of the third millennium B.C., there appeared the more abstract device of the *syllabary*, a collection of fixed symbols that refer not to objects but to syllables—to pieces of spoken language itself, as it were. This in turn yielded to the latest stage, with which we are still experimenting, namely the evolution of a fully phonetic alphabet with its immensely greater flexibility. Although the shift from the visual imagery of Egyptian, Mayan, or Chinese writing to an alphabet of abstracted symbols is typically depicted as evidence of intellectual progress, this linear view is belied by many features of our own literary environment:

The symbolic representation of the world on papyrus sheets, clay tablets, temple walls, or magnetic disks does not remove all traces of ambiguity (as proven by the fact that we still have poetry and arguments). But its invention revolutionized our means of apprehending that world. With the technologies of writing, the early civilizations accounted for their communal surplus, honored their gods, recorded the exploits of their kings, codified their laws, assessed taxes against one another, and developed a literature—one of whose preoccupations was, and still is, coming to terms with technology itself.

The earliest documents do not tell us much; mostly they are actuarial tables, inventories, and records of commercial transactions pressed with a wedge-shaped stylus (*cuneus*, hence cuneiform) into wet clay. "Writing," as J.D. Bernal wryly put it in his classic *Science in History*, "that greatest of human manual-intellectual inventions, gradually emerged from accountancy." By the time ancient scribes began to explore the literary possibilities of this new instrument, there was a great deal to account for—not only stores of private and communal property, but larger questions of origin, meaning, and purpose: How had they come to be there, in that place, subject to its gods and kings, and endowed with the tools, techniques, and traditions at their disposal?

Ancient answers to these questions tended to be considerably more dramatic than anything modern science has been able to propose. The oldest Egyptian cosmogony, or myth of origin, begins with the nocturnal copulation of Earth (Keb) and Sky (Nut); it is the daily task of the god Shu to separate them and lift over his head the goddess Nut so that her arched body may form the canopy of the sky. For the Maori of heavily forested New Zealand the same task is performed by Tu-matauenga, who forces apart the primal couple and holds the sky aloft with tall trees so that humans may have space to live. Beneath their mythic veils, these and a thousand other stories suggest a form of speculative thought. Sacred myths are not philosophy or science, but they lay out no less effectively a conception of nature in which humankind is embedded in a seamless web of interdependencies. The sacred technologies of the ancient engineer were designed to propitiate the forces that inhabited it.

One of the earliest written narratives is contained in the Akkadian-Babylonian hymn, the *Enuma Elish* ("When above"), which accounts for the origins of the universe in terms of a titanic struggle between the civilizing god Marduk and the primeval Tiamat, the monstrous female deity from whose moist body all things grow: "When above there was no heaven, no earth, no height, no depth, no name," the primordial universe was a vast watery expanse, inhospitable even to the gods. In a story that resonates with universal overtones, the younger gods band together and conspire to overthrow the powers of chaos. Marduk, mightiest of the younger generation, is sent out to confront Tiamat, the mother of all, embodiment of untamed Nature.

Where Tiamat has only "garbled spells" and "muttered maledictions" at her disposal, Lord Marduk brings to the battle the combination of supreme political authority and overwhelming technological force. He bears down upon her in his storm-chariot, with lightning glinting off his armor, a bow and quiver hanging at his side, his right hand grasping a mighty sword, his left carrying a huge net whose corners are held by the Four Winds. His battle plan is simple but daring. He first entangles the primal mother in his net, and when she opens her cavernous mouth to swallow him, sends in *Imhullu*, the Evil Wind, to hold it open. He then wounds her fatally with an arrow, and with his sword *Abubu*, "splits her like a shellfish into two parts." With the upper half he builds the arc of the sky, while the bottom becomes the platform of the earth.

This gruesome myth has been recognized as a primitive explanation for the ordering of the sky and the earth, for the family of gods, the institution of kingship, and much else. It can also be seen as an allegory for the constant struggle of mortals to restrain the encroaching waters of the Mesopotamian delta, and to reclaim a bit of silted ground on which to grow their crops and build their cities. The annual chanting of the hymn, on the fourth day of the New Year, re-enacted the original victory of order over chaos, and (it was hoped) secured the fruits of that victory for another year. With all of its mythical and magical overtones, it was as much a part of the technological apparatus of the Babylonians as were their quays and dikes and canals.

If the *Enuma Elish* served the ancient Babylonians as a technological artifact, it serves us as a textual artifact from which to reconstruct their views of technology. Marduk, it will be recalled, does not simply blow his mother up like a frog; he summons *Imhullu*, the Evil Wind, to do the job, and his sword

Abubu finishes her off. His tools, and even the forces of Nature, are conceived not as a 'what' but as a 'who.' They are alive and permeated with con-sciousness, and may be enlisted—given the proper ceremonies, rites, and rituals—as allies in the joint enterprise of survival.

One must tread carefully in such a world, for it is alive with intelligence, feeling, and will. It is a world in which the line between subject and object, animate and inanimate, has not been drawn, and in which one's allies may without notice turn into mortal adversaries. The engineer is thus a magician, a scholar, and a priest: The Egyptian King Cheops summons his magician to fold the sea back on itself in order to recover a pendant lost by a favorite wife, but also to perform more serious operations. It is not surprising that the Babylonian Code of Hammurabi (c. 1728–1626 B.C.) provides that anyone convicted of practicing magic without a license "shall go to the river and shall throw himself in"; his accuser may take away his house.

The primitive technologist, then, is midwife to a sexualized, invariably female nature, with the task not of innovating or inventing but of imitating, and thus hastening the delivery of the fruits of nature herself. He (and, rarely, she) may become a priestly official whose tools are magical and whose techniques are shrouded in ritual and mystery. The title given to the project manager of the earliest of the Egyptian pyramids captures some of the reverence accorded to the artisan-priest: "Chancellor of the King of Lower Egypt, First after the King of Upper Egypt, Administrator of the Great Palace, Hereditary Nobleman, High Priest of Heliopolis, Builder, Sculptor, and Maker-of-Vases-in-Chief, Imhotep."

From the earliest inscriptions, it is clear that the technical expert—not as master of nature but as its agent, arbiter, and accomplice—could attain to the highest status. Imhotep was venerated by Egyptians; his counterpart Asclepius was deified by the Greeks (in one of their many borrowings from Africa) as the god of healing. His descendants in Egypt and his counterparts in the traditional cultures of Asia, Europe, and the Americas were tinged with the divine because they knew how to coax favors out of their enchanted environments. Nearly five thousand years would pass before the reduction of nature to mere inanimate matter was complete; the process has been called "the disenchantment of the world."

Just as the forces of nature were seen as individual co-habitants of a single, continuous cosmos, so the creations of human technical ingenuity acquired individuality and autonomy in the culture of the ancient world. The author of the "Myth of Enlil and Ninlil," in describing the central Sumerian city of Nippur, captures just this continuum of the human, the artificial, and the natural:

> We are living in that very city, in Duranki,
> We are living in that very city, in Durgishimmar.
> This very river, the Idsalla, was its pure river,
> This very quay, the Kargeshtinna, was its quay,
> This very harbor, the Karusar, was its harbor,
> This very canal, the Pulal, was its well of sweet water,
> This very canal, the Nunbirdu, was its sparkling canal.

Although we moderns also give names to freeways, landfills, and data bases, the river and the canal of Nippur were not "named after" but simply *named*. The ancient poet must have felt secure among the animate artifacts with whom he dwelled, for he was clearly swollen with urban pride—over his city's neatly orthogonal streets and imposing temple square, its harbor and quay, and its freshwater canal (in which maidens are warned not to bathe unattended). While his attitude toward his surroundings remained reserved and respectful, the capacity of human beings to transform their environment had begun to register upon his consciousness.

It is not surprising that when the first fragments of writing appear toward the end of the fourth millennium B.C., the city itself—in all of its power and splendor, with its diverse and stratified population, its private homes and its public works—should emerge as a living landscape against which the encounter between human technological creativity and the forces of nature is played out. This confrontation is nowhere more vividly portrayed than in the *Epic of Gilgamesh*, which recounts the exploits of the historical king of Mesopotamian Uruk, who lived early in the third millennium B.C.

When we meet the figure of Gilgamesh he is a hero straining against the limits of his mortality, storming through town like a bull, leaving no son to his father and no daughter to her betrothed. In desperation his subjects conspire to tame him by confronting him with a force even wilder than he: Enkidu, his Upper Paleolithic alter ego, whose body was "shaggy with hair," who "jostled at the watering hole with the animals" and who "knew neither people nor settled living." Whereas Gilgamesh has overstepped the bounds of civility, the nature-man Enkidu has not yet attained it. Locked in combat they neutralize and civilize one another, and are prepared to rejoin the human city.

This is the critical passage in the history of civilization, and it is, as in so much of the world's mythology, watched over by various female characters who serve to warn against the unbridled conquest of nature: a temple prostitute, who initiates Enkidu into the urbane life of the citizen; Ishtar, the headstrong goddess of love, who threatens on the slightest provocation to turn the tiller of soil into a mole and the shepherd into a predatory wolf; the innkeeper Siduri, who stands on the banks of the River of Death to remind the traveler that he cannot transcend his own mortality and had better reconcile himself to the human city and a life spent peaceably within its walls.

The city itself, that most spectacular technological artifact of the ancient Near East, recedes into the background, but we may just as easily see it as a central character in the epic. At the heart of the story is the hero's journey outward in a quest for immortality and his humbling return. As we will find repeatedly, however, behind the return of the hero lie the nuts and bolts, the bricks and mortar, of technology. The narrative closes as it opens,

Figure 1.1

*And now they brought to them the weapons, they put in their hands the
great swords in their golden scabbards, and the bow and the quiver.
Gilgamesh took the axe, he slung the quiver over his shoulder, and the bow
of Anshan, and buckled the sword to his belt; and so they were armed and
ready for the journey.*

[The Epic of Gilgamesh]

The "Standard of Ur," third millennium B.C.

The upper row shows a column of Sumerian infantry, armed with
bronze-tipped spears and protected by helmets and body armor;
below are two four-wheeled war chariots.

with Gilgamesh standing astride the mighty double walls he has built for Uruk, gleaming like copper, intricately constructed of kiln-fired brick. He will fall silent and his mortal body will die and be laid in the dust, but the clay walls on which he has inscribed the tale of his exploits survive to confer upon him a belated immortality. Their remains, nearly six miles in perimeter and once punctuated by 800 semicircular towers, testify for all time that he learned not only civility, but civil engineering.

The peoples of antiquity had been building such monumental structures—"megamachines," Lewis Mumford called them, in recognition of the regimented human laborers who comprised their moving parts—long before the beginning of recorded history. The walls of Jericho, begun in the eighth millennium B.C., were four meters high and enclosed an area of 27 acres in which 3,000 people resided permanently; to build just the Pyramid of Cheops an army of Egyptian laborers had to quarry, cut, and transport 6 $\frac{1}{2}$ million tons of limestone; the 26-ton megaliths at Stonehenge are arranged with a degree of astronomical precision that continues to baffle investigators. Only in recent decades have these immense engineering structures begun to be truly understood, and the technological subtlety they reveal is often staggering. They are, for better or for worse, the most compelling evidence of the ancient commitment to civilization; it is almost as if the earliest town-dwellers felt driven to secure, under the weight of limestone blocks, sun-baked bricks, and granite obelisks, their fragile victory.

This hypothesis is attested by a cuneiform fragment, in which a Sumerian towndweller professed his superiority over a neighboring people "who do not know houses and do not cultivate wheat." By modern standards the author himself had not progressed very far beyond those nomadic wanderers to whom he referred, some 4,000 years ago, with such evident disdain, but in the telescopic vision of time, his society had made a quantum leap. It had potter's wheels and kilns in which artisans could bake fine pottery at 1,650° F.; it possessed a formidable military arsenal that included armored regiments of bowmen and javelin throwers, and royal armories to equip them; it could move water, stones, and human labor power over considerable distances; it was organized around a complex temple economy whose surplus supported specialized groups of stonemasons, carpenters, and metalsmiths able to beat swords into plowshares—and back again—with relative ease; its emergent division of labor was overseen by a corporation of priests and a hierarchy of scribes whose actuarial tablets thrust the human race into the adventure of literacy and, therefore, of history.

The decisive difference, however, was that in contrast to his migratory neighbors, the townsman remained throughout the year in one place—a distinction that marks the onset of civilization and is called the Neolithic Revolution. This relatively abrupt process of social transformation, well under way in various parts of the world some 10,000–12,000 years ago and now almost complete, introduced two new and apparently unprecedented practices—the seasonal cultivation of the land, and the permanent domestication of animals—and two corresponding human types, the farmer and the shepherd. Equipped with the hoe, the adze, and the sickle, the conditions of their lives were radically different from the hunters and gatherers whom they displaced. Rather than passively receive the means of his subsistence, the new professional had to provide for his family, as the Hebrew Bible puts it, "by the sweat of his brow."

The biblical expression is no mere rhetorical flourish. For the peoples of the Near East, the arduous work of civilization must have remained a central preoccupation, preserved in myth and literature centuries and even millennia after it had begun. Indeed, the Bible, that great national epic whose earliest books date back to the consolidation of the first Israelite monarchy around 1,000 B.C., provides some of the best indications of the technological apparatus the inhabitants of the ancient Near East brought to this task. The characteristic themes of Sumerian, Egyptian, and early Semitic literature come together fully in the Hebrew Bible, which offers the first hints of moral reservations about technology.

The Bible carries all the marks of the traumatic but comparatively recent passage out of nature and into cultivation, cult, and culture. In fact, the opening Book of Genesis, the grandiose story of the creation by God of the universe, is also the very mundane story of the creation, by humans themselves, of the earth as a dwelling-place for civilized men and women. From the security of the Garden of Eden, Adam and Eve were fomenting their own private neolithic revolution.

Eden was an oasis, "a garden eastward," bathed in sunlight and naturally irrigated by four alluvial rivers, wherein grew "every tree that was pleasing to the sight and good for food." So pristine was this original natural ecology that the herd animals of the field, the fowl of the sky, and the wild beasts of the forest had not even been harnessed with names. But "the Edenic impulse of name-giving," as Nietzsche called it, was only the first act in a drama that would be played out for the duration. Human beings, according to the two versions of creation laid side by side in the Book of Genesis, were given opposing mandates with respect to their natural surroundings. The oracular "Priestly" version authorizes them to "fill the earth and subdue it! Have dominion over the fish of the sea, the fowl of the heavens, and all living things that crawl about upon the earth." [Genesis 1:28] The author of the so-called "J" version, however, counsels prudence: The earth is not a workshop but a fragile garden, and the charge to human ingenuity is not to master, but "to till it and tend it." [Genesis 2:15] The ongoing task of civilization has been to resolve, reconcile, or maneuver between these two seemingly contradictory divine directives.

Nor do our biblical ancestors get off to a very promising start. Although generously provisioned with "every seed-bearing plant that is upon the earth" and every animal "in which there is the breath of life," the primal couple violate the express condition of their natural existence and are cast out. They have acquired the forbidden knowledge of morality and mortality; self-consciously, they have outfitted themselves in clothes. Within only one generation, their sons will have lost all trace of their hunter-gatherer ancestry as "Abel became a keeper of sheep, and Cain became a tiller of the soil." In the opening chapter of Western literature, the Neolithic Revolution is captured in the terrifying image of the expulsion from Paradise and the curse of labor (Adam) and labor pains (Eve). The very soil which had offered up its fruits is cursed, and the man is told to earn his living in the sweat of his face. Only their curiosity and ingenuity, which got them into so much trouble in the first place, will ease their burden.

Technological images abound in the Bible, but after this irredeemable catastrophe, it is not surprising that they rarely appear as great achievements. Noah's famous Ark, we are told, was made of gopher wood and sealed inside and out with pitch, a seaworthy 300 cubits long by fifty cubits wide by thirty cubits high, not to speak of its cabins and three decks. But this was no product of naval architecture, as Noah merely saw to its construction in compliance with a blueprint drawn in heaven. Nor did he then set out proudly to conquer the waves, for the deluge swept away all in its path, leaving the Ark bobbing helplessly like a cork until a greater power saw fit to deliver him to dry ground.

The biblical passage that best focuses the question of technology is the parable of the Tower of Babel, brilliantly compressed into the first nine lines of the eleventh chapter of Genesis:

> All the earth had the same language and the same words.
>
> And it came to pass, as they journeyed east, that they came upon a valley in the land of Shinar and settled there.
>
> They said to one another, "Come let us make bricks and burn them hard." Brick served them as stone, and bitumen served them as mortar.
>
> And they said, "Come let us build a city, and a tower with its top in the sky, to make a name for ourselves; else we shall be scattered all over the world."

The Lord came down to look at the city and the tower which they had built, and the Lord said, "If as one people with one language for all, this is how they have begun to act, *then nothing they propose to do will be out of their reach.*

"Let us, then, go down and confound their speech there, so that they shall not understand one another's speech."

Thus the Lord scattered them from there over the face of the whole earth; and they stopped building the city.

That is why it was called Babel, because there the Lord confounded the speech of the whole earth; and from there the Lord scattered them over the face of the whole earth.

This famous act of divine retribution, which expresses the intersection of technology, morality, and history, was more than just the imaginings of an inspired author. Beginning in 1854, with the excavations sponsored by British Consul J.E. Taylor at Ur, research began to reveal the forms of at least thirty stepped pyramids, or *ziggurats*, that dominated the Mesopotamian plane for perhaps 1,500 years prior to the decline of Babylon in the 6th century B.C. Between 1899 and 1917 the German archaeologist Robert Koldewey excavated one of them, the *Etemenanki* or "House of the Foundation of Heaven and Earth." This monumental terraced temple to Marduk (whom the Greek historian Herodotus rather presumptuously called "the Babylonian Zeus") proved to be the ancestor of the "accursed" tower of the Bible.

Accursed or not, the Ziggurat of Babel—like those of Ur-Nammu, Aqar Quf, Uruk, Nippur, and Eridu—were formidable structures indeed. The baked bricks that the biblical account says, sneeringly, "served them as stone" were perhaps the first industrially mass-produced commodities ever manufactured. Piled in inconceivable volumes, mortared with a carboniferous petroleum distillate the Bible calls "bitumen", and crowned with an imposing temple to the local deity, they rose in towers that commanded both the political landscape and the religious imagination. The foundations of the tower, claimed one late

Figure 1.2

If they can do this, then nothing that they may propose to do will be out of their reach.

[Genesis]

Pieter Bruegel the Elder, *The Tower of Babel* (1563)

cuneiform tablet, lay "secure in the bosom of the nether world," and the Bible confirms that its summit reached into the heavens: If they can do this, quails the Biblical narrative, "then nothing that they may propose to do will be out of their reach."

Since the early dynastic period the ziggurat had been the center of religious life in the Babylonian city, rising as a unifying symbol above its cosmopolitan "babble." To the monolatrous Jews, however, it was a symbol not of religion but of its repudiation, an assertion in stone of the presumptions of human technological prowess unbounded by the constraints of morality. The stepped tower, literally a staircase to heaven, expressed the will to stand on an equal footing with God; worse, its massive walls could be seen as a fortification to keep God out—out of the human city, out of the realm of human affairs. It was the most dramatic statement yet of the use of technology to delineate a boundary between the sacred and the profane, and nothing short of linguistic confusion could put an end to that ill-conceived adventure.

The first act of divine retribution had been the expulsion of Adam and Eve for aspiring to unlimited knowledge; What brought the wrath of God down upon the Babylonian tower-builders was the prideful application of their accumulated knowledge to an immense public works project. It may not be wholly fanciful to liken the authors of the Bible to ourselves, projected not so many years into the future, as we gaze upon our first permanent space stations: marveling at what we have made, imagining the possibilities they may open up, and wondering, given our record of violence and misjudgment, what perils they may portend. If the boldness of the ancient builders threatened to undermine the moral ground upon which civilization was built, then there were indeed grounds for divine consternation; for this was, as it were, just "the beginning." That spectacular, one-time feat—the creation of the heavens and the earth—has not yet been equalled, but it seems to be the destiny of the human race to try.

2

■ ■ ■ ■ ■ ■

Technology as Poetry:
The Classical Age

The Doric column, expressing a perfect

harmony of art, ideas, and engineering, raises one

of the most perplexing riddles of the ancient

world. "The Greeks and Romans built a high

civilization," marveled classical historian

Moses Finley, "full of power and intellect and beauty," with technologies inherited from their ancestors; but to their successors they bequeathed almost nothing new. Indeed, we can count on the fingers of one hand the number of technological innovations introduced by the people whose gifts to posterity include tragic drama, democratic self-governance, systematic philosophy, and the classical orders of architecture. Did they simply disdain the technical arts in favor of the life of pure reason? Or were they striving to contain their technology within a framework of cultural values and social institutions?

Starting in the last century, with Heinrich Schliemann's romantic quest for the ruins of Troy, archaeologists have largely confirmed the Homeric picture of the Greek Bronze Age. Beneath the surface of the *Iliad* and the *Odyssey*, the epic sagas of the siege of Troy and the wanderings of Odysseus, lies a veritable encyclopedia of information about the archaic household economy, early Greek customs and rituals, the protocols of warfare, and the pantheon of the gods. Nor has the technology of the Greeks been neglected: Though Helen's face may have been enough to *launch* a thousand ships, it took shipwrights, sailmakers, carpenters, navigators, and a crew of oarsmen to get this armada across the Aegean Sea to Troy. Increasingly the nature of Greek technology has come to light; the Greek *attitude* toward technology, however, remains shrouded in controversy.

Writing probably in the 8th century B.C., the poet we call "Homer" paints a vivid, if occasionally confused, picture of the epic world of proud Achilles and resourceful Odysseus, of the faithless Helen and the exemplary Penelope. The confusion arises from the fact that Homer was drawing upon a collective memory of events that would have transpired in the distant past, and he often mixes details of his own "Archaic" world with themes drawn from the vanished Mycenaean civilization whose warriors had besieged the walls of Troy some 400 years before.

Although usually described as the story of the Trojan War, the *Iliad* is actually, as the author himself informs us, the song of a goddess about the prideful anger of Achilles and the devastation that follows in its train. The human world of armorers, weavers, shipbuilders, farmers, potters, tanners, and carpenters whom we meet along the way is thus set against a none-too-reverential family portrait of the immortals whose pride and foolishness they mimic. Indeed, the story really begins in a family squabble among the gods.

While battle lines are being drawn on earth, our attention is diverted to Olympus, where the patriarchal Zeus has found himself caught once again between a favored nymph and his willful consort Hera. Into this scene of marital discord limps one of the strangest and least-appreciated gods of ancient Greece: their son Hephaestus, "the renowned smith of the strong arms," master of the forge and god of fire. In a spirit of strict technical objectivity he pleads eloquently for a reconciliation between his parents, "but among the blessed immortals uncontrollable laughter went up as they saw Hephaestus bustling about the palace." [*Iliad*, I: 599–600]

The deformed figure of Hephaestus, shuffling about among the perfectly formed immortals on his golden crutches, the butt of their jokes and object of their derisive laughter, affords a compelling insight into the second-class citizenship of the craftsman both in heaven and on earth. His misfortunes began in infancy when Hera, disgusted by the lame and sickly child she had born, dropped him into the sea. There he was sheltered by a pair of benevolent nymphs who set him up in an underwater cave with a forge with which he fashioned for them all sorts of useful and ornamental trinkets, including, it would appear, the world's first safety pin.

Impressed by his art (and envious of the fine brooches sported by his underwater patrons), Hera relented and restored Hephaestus to his place among the gods. He was given a veritable industrial empire of twenty forges and a robotic assembly line of golden automata, "in appearance like living young women."

And in spite of his disfigurement, the god of the forge was permitted to marry "sweet-garlanded Aphrodite" in an allegorical marriage, made in heaven, of technology and beauty. [*Iliad*, 18: 368ff]

One legend claims that despite his divine parentage, Hephaestus was "born without an act of love," as if to warn us that with his return to his family his troubles were not over. Indeed, he soon had the bad judgment to intercede on behalf of his mother in one of her never-ending conflicts with Zeus; now it was his enraged father who caught him by the foot and flung him a second time out of heaven, and this time he was a whole day falling. When the god of the forge finally crashed into the earth near volcanic Lemnos (hence his Latin name, Vulcan), he allows, "there was not much life left in me." Being one of the immortals, he could not die, but his legs were broken and he had to fashion a pair of golden crutches on which he hobbled about for eternity.

Thus we have in Hephaestus an incongruous and slightly ludicrous figure. He is lame, sweaty, and ill-tempered, with "massive neck and hairy chest," a proletarian god among the idle rich whose nectar and ambrosia is drunk from exquisite goblets which he himself has cast. But "the glorious smith of the strong arms" compensates for his own physical imperfections through the products of his workshop, which are flawless emblems of the marriage of art and industry. A bard in the *Odyssey* tells how, on one unhappy occasion, Hephaestus learns that laughing Aphrodite is lounging in his own marriage bed with the warlike Ares:

> Hephaestus, when he had heard the heartsore story of it,
> went on his way to his smithy, heart turbulent with hard sorrows,
> and set the great anvil upon its stand, and hammered out fastenings
> that could not be slipped or broken, to hold them in fixed position.
> [*Odyssey*, 8: 266ff]

As the faithless lovers slept, "all about them were bending the artful bonds that had been forged by subtle Hephaestus," so that they awoke to find themselves prisoners of their adulterous passion. Thus the divine craftsman exacted his revenge, but not even his consummate technical skill could erase the stigma of his lowly trade.

Here, indeed, is the tragic irony of the limping god of technology, for while he is deceived by his wife and despised by his fellow immortals, they do not hesitate to call upon his technical services: It is Hephaestus who designs the armor of Achilles, inlaid in gold and silver with the history of the Greeks; who crafts the shining helmet of Athena and the scepter of the great king Agamemnon; who is later summoned to forge the unbreakable chains that will bind Prometheus to his rock. And it is Hephaestus who is charged to fashion the womanly Pandora, "gift of all the gods" to mankind. Clearly, however, it is the products, not the producer, that are valued. Though he is honored with torch-races and a temple atop the Acropolis, "the god of the dragging feet" provides the Greek pantheon with a rich source of irreverent comic relief: a popular depiction on vases shows Hephaestus being escorted back up to Mount Olympus, disgracefully inebriated, seated backward on a sexually aroused donkey.

It is no coincidence that around the Hephaesteion, the monumental temple that stands alone on a hilltop west of the Athenian Agora, archaeologists have found numerous bronze-casting pits and other evidence of industrial activity—as if to suggest that the god of the forge watched over the low-born mortals who labored nearby. With his powerful forearms and prominent limp, the village smith is, in fact, a familiar figure not only in Greek mythology but in folkloric traditions as diverse as those of Scandinavia and West Africa and in the Vedic poetry of India. The lame smith is a despised but valued outsider, able to make or mend the tools required by his fellow citizens but bound to his menial trade by an infirmity that disqualifies him from using them. On earth or in heaven, the master of the forge sustains the community in higher pursuits—warfare, farming, hunting, athletic competition—in which he himself is unable to

Figure 2.1

Sing, clear-voiced Muse, of Hephaestus, famed for his inventions.
With grey-eyed Athena he taught men glorious crafts.

Homeric hymn to Hephaestus

Heads of Hephaestus and Athena in the workshop of a smith, c. 480 B.C.

participate. He is ill-tempered, anti social, and slightly suspect, and his skill is tinged with sorcery. There is something plainly disreputable about him.

The Mycenaean civilization evoked in the Homeric epics, roughly 1600 to 1150 B.C., was based on a technology of bronze, an alloy of copper and tin that is durable, workable, and easily repaired (so long as one has access to a stone furnace capable of smelting ores at temperatures above a modest 2,000° F.). Bronze had been around for a long time, and from it the peoples of the eastern Mediterranean had made their tools, their weapons, their ornaments.

Much of what we know of Bronze Age Greece comes from the indefatiguable labors of Heinrich Schliemann, the amateur archaeologist who knew the *Iliad* and the *Odyssey* by heart and who in 1876 excavated the fortified citadel at Mycenae. Schliemann may not, as he excitedly telegraphed the King of Greece, have "gazed upon the face of Agamemnon," but he did discover the material remains—the massive architecture, the tools and weapons, the artworks, and the luxuries (lapis lazuli, ostrich eggs, linen, crystal, and amber)—of a warrior aristocracy that abruptly disappeared sometime around the presumed date of the Trojan War. During the catastrophic centuries that followed, the Greeks ceased to build cities; they lost the script known to archaeologists as "linear B"; they lost the arts of working in stone, painting frescoes, and carving ivory. Most importantly, perhaps, they lost either the ability or the resources to work in bronze.

When cultural activity resumed in the Greek-speaking world, it was marked by a dramatic increase in the use of iron. Iron had also been around for a long time but was never of much interest; in the crude stone furnaces of antiquity it reduced to a soft, spongy mass of no practical value. Further, molten bronze could be cast in reusable molds while iron implements had to be individually forged; bronze corrodes slowly and develops an attractive greenish patina, whereas iron quickly rusts; and the tensile strength of worked bronze is triple that of the porous "bloomery iron" known to ancient metallurgists.

The breakthrough came when it was discovered (empirically, of course) that smelting iron over white-hot charcoal diffuses carbon atoms throughout the ore; this produces a primitive form of steel whose tensile strength, when quenched in water and hammered, leaps to more than twice that of bronze. Homer mistakenly assumes that his Bronze Age heroes were as familiar with this process as his Iron Age audience four centuries later: When Odysseus blinds Polyphemus the Cyclops with a sharpened stake, it is "as when a man who works as a blacksmith plunges into cold water a great axe blade which hisses aloud, tempering it, since this is the way that steel is made strong." [*Odyssey*, 8: 266ff]

There is evidence of a steady increase in prosperity around 900 B.C. that foreshadowed the Greek renaissance crowned by the *Iliad* and the *Odyssey*: It is marked by the adaptation of a Phoenician alphabet, the institution of the Olympic Games, the revival of long-distance trade, and by the widespread use of iron. This is the world of the dour and law-abiding Hesiod, a poet roughly contemporary with Homer but inhabitant of a different mental universe. Hesiod, a déclassé gentleman-farmer of Boeotia in central Greece, poured out his sorrows in a group of poems that are saturated with mythological and technological detail. But where Homer's gaze was backward—across the Greek Dark Ages to the gilded Mycenaean kingdoms of Agamemnon, Achilles, and Odysseus—the hard-headed, no-nonsense Hesiod was planted firmly in the mud and muck of his own fields.

The two poets are separated by more than differences of style and temperament, however, since nothing less than a technological revolution divides the eras they evoke in their poems. Hesiod's *Works and Days*, a sort of farmer's almanac of homey wisdom, records the passage. It begins by recalling how the gods created first a perfect race of golden mortals who lived among them, then a silver age of idlers, a warlike race of bronze, and a mysterious race of heroes. He concludes his metallurgical metaphor with a grim picture of "the fifth, who live now on this fertile earth":

> I wish I were not of this race, that I
> had died before, or had not yet been born.
> *This is the race of iron.* Now, by day,
> Men work and grieve unceasingly; by night
> They waste away and die.
>
> [*Works and Days*, 174–8]

In this famous myth of metals Hesiod escorts us out of the Stone Age and into history. Most striking, however, is the implication of technology—for moderns the most pervasive symbol of progress—in what is plainly a cycle of decline. For Hesiod mortals have become successively more tarnished, corroded, and rusted out; he surely does not believe that technological progress has relieved the human condition.

Hesiod counsels his readers to avoid get-rich-quick schemes involving trade, seafaring, or other labor-saving shortcuts: "If in your heart you pray for riches," he admonishes his ne'er-do-well brother Perses, "pile work upon work and still more work." To explain this dull air of resignation we must turn to his genealogy of the gods that watch over the hapless mortals of his day: Presiding over the industrial revolution of which he himself grudgingly partakes is the crafty trickster Prometheus, whose crime and punishment expose the hazards of the technological fix.

Hesiod's Prometheus, literally the "fore-seer," was of the oldest generation of gods, older even than the upstart Zeus who had by this time gained dominance in the Greek pantheon. "The gods desire to keep the stuff of life hidden from us," Hesiod complains, but Prometheus, in an act of cosmic defiance, appointed himself champion and benefactor of the maligned race of mortals. Everyone knows how he stole from Zeus a spark of fire, symbol both of enlightenment and destruction, "concealed the flame in a fennel stalk, and fooled the Thunderer."

Figure 2.2

Prometheus, most crafty god of all,
You are happy that you stole the fire and tricked me
But it will be a great sorrow to you
And to men who come after.

Hesiod, *Works and Days*

Laconian Cup depicting punishments of Atlas and Prometheus

The theft of fire became a permanent theme of Greek mythology, developed nowhere more powerfully than in *Prometheus Bound*, the first part of the dramatic trilogy by the 5th-century playwright Aeschylus. In his hands Promethean Fire takes on a deeper meaning than it did for Hesiod, for it is now bound up with instruction in architecture, astronomy, mathematics, writing, navigation, medicine, metallurgy, and the interpretation of dreams. "All human skill and science was Prometheus' gift," boasts the god; "I saved the human race from being ground to dust." [*Prometheus*, 442]

In the version left by the sullen Hesiod, the human race is punished for receiving the illicit gift of fire, as Pandora opens her jug and looses upon them labor, sickness, and all other misfortunes. In Aeschylus, however, it is rather the defiant bringer of technology who is subjected to horrendous punishment. In either case, it is clear that enlightenment is not an unalloyed gift, and the other figures of Greek mythology associated with technology merely confirm this general reserve. Hermes, who invents the alphabet, boxing, and weights and measures (all before he was two days old!), is a thief; Daedalus, the technological genius who invents carpentry, the folding-chair, and statues that could walk and swing their arms, knows that we must not overstep the bounds of our technological skill: "My son, be warned!" he instructs young Icarus, upon crafting for him the wings with which they will escape the Cretan Minotaur: "Neither soar too high, lest the sun melt the wax; nor swoop too low, lest the feathers be wetted by the sea." His warning goes unheeded, however, and Icarus, intoxicated by the emancipation of flight, flies toward the sun and crashes into the sea. Ultimately, the Minotaur will be defeated when Theseus ventures into its Labyrinth, trailing behind him a ball of string provided by the admiring Ariadne. It can be no coincidence that victory owes more to the ingenuity of the Princess than to the arms of the Hero.

Why would the mortal Greeks, recipients of "the flowery splendor of all-fashioning fire," so humble the gods of technology? Why would they permit their own most loyal benefactor, Prometheus, to be chained to a rock in the Caucasus where every day for 40,000 years a vulture came and gouged out his immortal liver? Could it be that they suspected that they deserved the limits Zeus had imposed upon them? Could Aeschylus have perceived that one gift was missing from the polytechnical education imparted by Prometheus, namely the wisdom that would enable an immature humanity to manage its technological powers responsibly?

But more than that," says Prometheus brushing aside the lesser charges arrayed against him, "I gave them fire." If we update the symbol of fire to "fissionable plutonium" we may hear a distant echo of our modern fear that we have grown too clever for our own good. Technology, the Greeks seem to have been saying, is an indispensable part of culture, but it must be used in the service of culture.

The dramatic accomplishment of Aeschylus ushers us into the period of Greek civilization that has captured the imagination of posterity. The philosophical foundations of the Classical Age were laid by the bold intellectual experiments of the so-called "Pre-Socratic" philosophers—Thales, Anaxagoras, and Heraclitus of Ionia; Pythagoras, Parmenides, and Zeno of the Italian schools; and a handful of others. Though themselves indebted to much older currents of Egyptian, Hittite, and Babylonian thought, and known to us now only by the merest fragments, they represent some of the earliest attempts to derive a naturalistic rather than mythological understanding of the universe. Since Plato their speculations have been ridiculed—Thales presumed that the essential principle of the universe was moisture; for Anaximenes it was air in different states; Democritus conceived the outlandish idea that matter was composed of tiny spheres which he called by the Greek word *atomoi,* meaning "indivisible." But by asking of what is the universe constructed? what is the cause of change? what are the origins of life? they helped to stimulate the demythologizing of thought that continues, unfinished, into our own time.

The innovation of the classical philosophers—Socrates, Plato, and Aristotle—was to thrust the human being, as individual and as member of the political community, into the picture. Since humans are tool-making and tool-using beings, and since the classical Greek city-state was an economic as well as a political community, it was inevitable that technological themes would find their way into their philosophic discourse alongside the search for the Good, the Beautiful, and the True.

We know that the aristocratic Plato was familiar with the technical trades of his day, for his dialogues are peppered with similes drawn from the potter's wheel, the blacksmith's forge, and the carpenter's bench; one tradition holds that he even invented a water-actuated clock to signal the beginning of classes. The real clue to Plato's views about technology, however, lies buried in the structure of his philosophy as a whole.

At the heart of Plato's thought lies the idea that the highest form of reality resides not in the "works and days" that had preoccupied Hesiod but in an ideal realm of essences or pure and incorruptible Forms. The effect of this doctrine is to redirect our attention away from the arts and crafts, the nuts and bolts of everyday life, to a higher sphere beyond the ravages of time. It follows that the artisan can be no true creator, for he is capable of working only with tangible objects (beds and chairs, significantly, are the philosopher's favorite examples!) and not with the Ideas of which his products are but crude and imperfect approximations. Only in one sense is the mechanic able to attain to a pure Form, and that is through "the surfaces and solids which a lathe, or a craftsman's rule and square, produces from the straight and the round." It is not his creative artistry that is celebrated here, but his capacity to approximate forms that are "always already" beautiful, "in their very nature." [*Philebus,* 51c]

Needless to say, the place of honor in Plato's Republic is reserved for the philosopher, who dwells wholly within the realm of Ideas and alone has access to it. The technical arts have a place in his grand scheme, for they are essential to the physical survival of the community, and Plato is respectful of those who master them. But he insists that each individual practice only the one art "for which he is fitted by nature," [*Republic,* II: 374b] and builds an entire speculative commonwealth around the idea of a strict and explicitly hierarchical division of labor: Just as the rational mind rules the base physical appetites of the body, so the body politic as a whole must be ruled from the brain of the philosopher. "Citizens," declares Socrates, "you are brothers, yet God has framed you differently. Some of you have the power of command, and these he has composed of gold, wherefore also they have the greatest honor; others of silver to be auxiliaries; others again who are to be farmers and craftsmen he has made of brass and iron."[*Republic,* III: 415a]

The division of labor, however hierarchical, that Plato advocates in *The Republic* ensures for the technical crafts a place in the political community. But in the commonwealth of his contemporary, Xenophon, even this is cast into doubt. Around 362 B.C. Xenophon published a philosophical dialogue in which an invented Socrates explains why the mechanical arts have acquired a bad reputation: "They utterly ruin the bodies of those who work at them," he observes, "by compelling them to remain indoors and in some cases to spend the whole day by the fire." But this is only the beginning, for "this physical degeneration results also in deterioration of the soul. Furthermore, the workers at these trades lack the leisure to discharge the responsibilities of friendship and of citizenship." [*Oeconomicus,* IV: 2–13] Technology, in Xenophon's treatise, has become the antithesis of citizenship, and he notes that some cities have gone so far as to make it illegal for a citizen to pursue a mechanical trade.

The last classical thinker to address himself to the politics of technology is Aristotle, Plato's student and successor, who devotes an important chapter to the question of whether technical tradesmen may be granted citizenship and share in the offices and honors of the state. "The truth is that we cannot include as citizens all who are 'necessary conditions' of the state's existence," he concludes, "for no man can practice virtue who is living the life of a mechanic or laborer." [*Politics,* III. 5: 1278a]

The hierarchy of head over hand, mind over body, theory over practice, and science over technology was deeply rooted in the thought of the classical age. Nowhere, however, is it more explicit than in the philosophy of Aristotle, whose passion for classification extended not only to species of plants, genres of poetry, and forms of governance but to knowledge as such. *Theoria*, pure theoretical knowledge, is of the highest rank, since it deals with the fixed and unalterable laws of the physical universe and can describe them with absolute precision. Mathematics, physics, and the special science of "meta-physics" belong to the sciences of "things that are true by nature," whose goal is knowledge itself.

In contrast to theoretical knowledge, the objects of the practical sciences are those things that comprise the world of human affairs and that are true not by nature but by convention— that is, things "which could be otherwise." The goal of these sciences is not so much a kind of knowing as a way of being: The ultimate object of the study of medicine is not medical knowledge but health; the purpose of political science, likewise, is justice. Once again, there is no doubt that the application of "science" to some crass utilitarian end is a necessary but inferior activity. Indeed, the authenticity of an ancient treatise on mechanics, once attributed to Aristotle, has been doubted precisely on the grounds of its author's passionate interest in the practical problems involved.

Aristotle introduces a further distinction when he implies that activity can be either practical or productive. Practical activity (*praxis*) is complete in and of itself; it is performed, one might say, for the process rather than the product, and might include philosophical disputation, athletic competition, or scientific speculation. By contrast, productive activity (*poiesis*) aims to produce some object: a meal, an amphora, a house. It hardly qualifies as knowledge at all for Aristotle, for it is devoid of inquiry into underlying causes, and the manual workers who perform it "are like certain lifeless things which act indeed, but act without knowing what they do." At first the inventors of these technical arts were admired, he opines,

> But as more arts were invented, and some were directed to the
> necessities of life, others to recreation, the inventors of the latter were
> naturally always regarded as wiser than the inventors of the former,
> because their branches of knowledge did not aim at utility.
>
> [*Metaphysics*, I: 981b]

Aristotle constructs an explicitly hierarchical system, in which action for its own sake is ranked above the technical production of useful things. Many such productive arts have now come into being, he acknowledges, "but to dwell long upon them would be in poor taste." [*Politics*, I. 11: 1258b]

Samuel Taylor Coleridge once reckoned that "everyone is born either a Platonist or an Aristotelian." On the question of technology, however, the representative philosophers of Classical Athens were entirely of one mind: The technical arts, insofar as they are directed toward some external necessity, are the proper business of slaves.

A couple of observations need to be made before we leave the Athenian city-state and its technophobic philosophers behind us. First, the disdain for technical skill was by no means universal among the writers of ancient Greece. The Hippocratic physicians believed that "where the love of mankind is there is the love of technique" [*Precepts*, VI] and concerned themselves as much with medical practice as with theory. An appreciation for technology is apparent likewise in the digressions of the historian Herodotus: "I have dwelt longer upon the history of Samos than I should otherwise have done," he explains,

> because they are responsible for three of the greatest building and
> engineering feats in the Greek world: The first is a tunnel nearly a
> mile long, eight feet wide and eight feet high, driven clean through
> the base of a hill nine hundred feet in height. . . . Second there is the
> artificial harbor enclosed by a breakwater, which runs out into
> twenty fathoms of water and has a total length of over a quarter of a
> mile; and last, the island has the biggest of all known Greek temples.
>
> [*Histories*, III: 60]

Though he tells a story of conspiracy, treachery, and murder, "these three works seem to me sufficiently important to justify a rather full account of the history of the island." Herodotus readily acknowledges, however, that in the estimation of his fellow Greeks, "craftsmen and their descendants rank lower in the social scale than people who have no connection with manual work." [*Histories*, II: 167]

Second, the philosopher had better reason than the doctor or the historian to be suspicious of technology. The mechanical arts are seen as debasing, Plato reasons, because those who practice them are incapable of ruling themselves. [*Republic*, IX: 590c] Indeed, he could hardly have concluded otherwise, for by the 5th century B.C. the majority of mechanical, industrial, and agricultural work in the Greek city-states had come to be performed by slaves who were, quite literally, "incapable of ruling themselves." The technical, mechanical arts, performed by unfree workers, were thus associated with slavery *per se*, while a contemplative ideal of the free citizen emerged in contrast to it.

Although the philosophy of the Greeks appears frankly reactionary with regard to technology, there lies buried in the Greek language a startling insight: the same word, *poiesis*, describes the work of both the mechanic and the poet. In modern times we have grown accustomed to thinking of the inspired artist and the disciplined engineer as opposed human types who gaze upon one another with mutual incomprehension, if not outright and avowed hostility. The Greeks, however, could not even express in language the difference between the artist and the engineer. Had either of them been admitted to Plato's Academy there would have been no quadrangle to keep them apart.

Plato and Aristotle wrote their treatises on the proper human commonwealth not because it was flourishing, but precisely to arrest its decline. The face-to-face democracy that united the adult male citizens of the Greek city-state never recovered from the Peloponnesian War, and the Philosopher-

King theorized by Plato gave way to the imperial conqueror—
Alexander of Macedon—actually tutored by Aristotle. In 334
B.C. the Alexander's armies crossed the Hellespont; by the
time of his death in 323, his empire sprawled across three
continents.

The Hellenistic empire or *cosmopolis* (literally a political
community that embraces the whole of the known universe)
spelled the demise of the democratic city-state. At the same
time, the integration of a vast geographical region into a single
cultural unit proved to be a massive stimulus to the growth of
both scientific and technological knowledge, and the old
aristocratic distinction between "applied" technology and "pure"
science began to blur. Athens declined as a center of learning,
supplanted by the opulent city of Alexandria. Just as the
philosophy and art of the Hellenistic era descended from the
Platonic realm of pure forms to a fascination with realism and
the conditions of everyday life, science and scholarship likewise
began to be grounded in practical applications. From the 3rd
century B.C., under the patronage of the Egyptian Ptolemies,
Alexandria become the site of a burst of technological creativity
wholly without precedent in antiquity.

The military feats of Alexander the Great can be credited in
no small measure to the engineers who built the roads along
which his armies marched, arranged for their logistical support,
and constructed the siege towers and torsion catapults with
which they battered down the walls of cities standing in the path
of his advance. The siege of the fortified island-city of Tyre in 332
B.C. was made possible not only by the discipline, courage, and
ruthlessness of Alexander's troops, but also by the technical
experts who oversaw the building of a massive stone breakwater
to connect the city to the Phoenician mainland; Alexander himself,
as the Roman historian Arrian informs us, "was always on the
spot giving precise instructions as to how to proceed, with many
a word of encouragement and special rewards for conspicuously
good work." [*Arrian*, II: 18] The haunting couplet of Sappho, the
mysterious 6th-century B.C. poet of Lesbos, anticipates the

military-industrial complex perfected centuries later by the youthful Alexander:

> Then the god of war, Ares,
> boasted to us that he could haul off
> Hephaestus, master of
> the Forges, by sheer force.

> [Sappho, 89]

Throughout the eleven-year campaign of Alexander, northward from Egypt, across Persia, and over the Khyber Pass into India, the servile "master of the forges" was again and again pressed into the service of the warlike Ares.

The Macedonian victors, beholden to their military engineers, did not forget their debts to science and technology during the imperial peace that followed. Whereas Plato's Academy and the Lyceum of Aristotle had been built around the personality of a single master, in Alexandria the wealth of an empire was channeled into the greatest state-supported institutions of higher learning to be found anywhere in the ancient world. The city's famous Library and Museum, endowed by the Macedonian rulers of Egypt, flourished for almost seven centuries as the most important centers of learning in the ancient world.

Scholars from across the Mediterranean, Persian, and Babylonian world mingled in the academy of Alexandria, with its manuscript collection of perhaps 700,000 scrolls, its collection of scientific instruments, its observatory, dissection rooms, botanical gardens, and zoo, and—not least important—the subsidies it provided to the scholars themselves. Euclid completed his systematization of Greek geometry there between 320 and 260 B.C.; Aristarchos of Samos, another member of this remarkable community of scholars, worked out a consistent heliocentric theory that would be revived some 1,800 years later by Copernicus; from Alexandria, Eratosthenes calculated the circumference of the earth and, with the same attention to naturalistic detail that is the defining feature of Hellenistic art, composed his monumental *Geography* around 235. Centuries later the astronomer Ptolemy worked in Alexandria's observatory; Galen completed his medical studies there. This remarkable episode ended when around 415 A.D. the Neo-Platonist mathematician Hypatia was stoned to death by a mob of Christian monks who burned the Library to the ground in their crusade against the pagan learning she embodied.

Hellenistic mechanics has also earned a place in the history of science: Its main figures are Ctesibios of Alexandria (c. 270 B.C.), Philo of Byzantium (c. 200 B.C.), and the inventor Hero, who worked in Alexandria in the 1st century A.D. In addition to inventing pumps, water clocks, pipe organs, fire engines, automata, crossbows, surveying instruments, and other mechanical contrivances, each of them wrote theoretical treatises on technical subjects. Their writings do not, however, cast much light on the value assigned to mechanical technology in the Hellenistic culture, but an account of the most famous of them all—the 3rd-century B.C. mathematician, Archimedes of Syracuse—suggests a clue. Although he is credited with the water-raising "Archimedean Screw," the compound pulley with which he boasted to King Hiero that given a place to stand he could move the Earth, and various other mechanical inventions, he seems to have regarded them as no more than "mere accessories of a geometry practiced for amusement," and wrote only about spirals, the theory of floating bodies, the geometry of the sphere and the cylinder, and other essentially theoretical problems. "He would not consent to leave behind him any treatise on invention," speculates his biographer Plutarch, "for he regarded the work of the engineer and every art that ministers to the needs of life as ignoble and vulgar." While it is true that he tinkered with engines of war and of peace, Plutarch continues, Archimedes' "lofty soul" would permit him to devote his serious efforts "only to those studies whose subtlety and charm are not affected by the claims of necessity." [*Life of Marcellus*, V: 17]

It is likely that Plutarch—a dyed-in-the-wool Platonist—is telling us as much about himself as about Archimedes, and perhaps much more. Plutarch (50?–120? A.D.) is best known for his *Parallel Lives* of notable Greek and Roman personalities: Alexander and Caesar were paired as imperial conquerors, Theseus and Romulus as legendary founders of states. In this spirit, we may compare the philosophical remove evident in Greek writing about technology with what we could call, with the pun very much intended, the more "concrete" attitude prevalent among the Romans.

Plutarch's disdain for manual labor and for the application of science to practical tasks was widespread among the aristocratic classes of Roman society. "Others will cast more tenderly in bronze," allows Vergil, the spokesman for the Imperial consciousness; but his countrymen need not lament their lack of technological prowess, for their destiny lies elsewhere, as he reminds them in the *Aeneid*:

> Roman, remember by your strength to rule
> Earth's peoples—for your arts are to be these:
> To pacify, to impose the rule of law,
> To spare the conquered, battle down the proud.
>
> [*Aeneid*, 6: 848–57]

Posterity has evaluated the Roman accomplishment rather differently, however. Although Roman legions did succeed in subduing the Mediterranean world for four centuries, they did so by deploying some of the most-sophisticated engineering principles of antiquity. Moreover, the greatest triumphs of Roman technology were in the field of civil engineering—directed, as that term suggests, to the *civitas*, to public works and the public good.

Except for eccentrics such as Pliny the Elder, few Latin thinkers regarded technical subjects as a field "in which the mind is eager to range." [*Hist. Nat.*, Preface] One Roman writer, however, brilliantly captures the public ethos informing the best of technologies: Marcus Vitruvius Pollio was not a poet or a philosopher but an *architecton*, a master builder, mechanic, and organizer of technical works. His treatise, addressed to the new Emperor Octavian (63 B.C.–14 A.D.) and known to us now as *The Ten Books on Architecture*, was the single most important body of architectural writings for 1,500 years, both because of the wealth of technical information it provided and because of the ethic and the aesthetic that infused it.

Although he dedicated his treatise to the ruling Imperator Caesar, the theories advanced by Vitruvius flew in the face of the slave society in which he lived and whose official poets, philosophers, and historians still doubted that the technical arts were compatible with the life of the free citizen. For Vitruvius, however, "knowledge is the child of practice *and* theory." [*Architecture*, I: 1] The integration of head and hand upon which he insists leads him to a vision of technology that includes structural integrity and mechanical efficiency, to be sure, but also the well-being of the whole citizen and the citizenry. Our present preoccupation with "human factors" would have shocked and horrified him—What is this larger thing of which humans are merely "factors?" he would have demanded.

In the opening sections of his treatise, Vitruvius describes the proper education of an architect. We encounter here a rather unusual curriculum—or so it must seem to us: Vitruvius requires that the architect not only be a competent draftsman and acquainted with the properties of different building materials, but trained also in history, philosophy, music, anatomy, law, economics, medicine, and a variety of mathematical disciplines including geometry, optics, and astronomy. Since the field is so vast, he concludes,

> I think that people have no right to profess themselves architects hastily, without having climbed from childhood the steps of these studies and thus, nursed by the knowledge of many arts and sciences, having reached the heights of the holy ground of architecture.
>
> [*Architecture*, I: 11]

Figure 2.3

Since nature has designed the human body so that its members are duly proportioned to the frame of the whole, it appears that the ancients had good reason for their rule, that in perfect buildings the different members must be in exact symmetrical relations to the whole general scheme.

Vitruvius, *De Architectura*

Carytides, from the edition of Vitruvius by Fra Giocondo, Venice, 1511

Architects must be trained in history and philosophy, for how else can their buildings be fitted into the ongoing life of the community? How else can they attain the values of dignity, high-mindedness, and that sense of justice without which "no work can be rightly done?" Likewise Vitruvius stresses that music must be included in the curriculum of the architect, for harmony is visual as well as accoustical. (Musical training is also helpful, he adds, in "tuning" the twisted sinews of catapults. Every machinist today would agree that the ear is as vital an organ as the eye, the nose, and the hand.) Law and medicine, too, are essential, because buildings are ultimately for people (though a few may be set aside for the gods) who may be adversely affected by the unhealthfulness of a site or by the lawsuits of their neighbors. And it is unimaginable to Vitruvius that an architect not also be skilled in geometry and optics, for just as nature has designed the members of the human body in proportion to the whole, so in perfect buildings the different members must be in exact symmetrical relations to the whole general scheme. [*Architecture*, II: 4]

Throughout his treatise Vitruvius discourses freely upon a variety of technical subjects: concrete and columns, vaults and arches, cranes and ladders, farmhouses and frescoes, sundials and waterclocks, and assorted bits of military and industrial hardware. In the last analysis, however, he insists upon the subordination of technology to wisdom, judgment, and prudent planning. He concludes with an account of the siege of Rhodes by the Athenian king Demetrius, who approached the city with an immense *helepolis*, a siege tower 135 feet high and 60 feet broad and capable of deflecting a 360-lb. stone shot from a catapult. The panicked citizens appealed to the architect Diognetus who, more crafty than proud, saw that he was outgunned and resorted to what we can only call the Scatological Defense. All through the night, as the attackers inched the tower closer and closer to the city walls, the citizens of Rhodes emptied their sewage through a breach he had made in the wall:

After a great amount of water, filth, and excrement had been poured out during the night, on the next day the helepolis moving up, before it could reach the wall, came to a stop in the swamp made by the moisture, and could not be moved forwards, nor later even backwards. And so Demetrius, when he saw that he had been baffled by the wisdom of Diognetus, withdrew his fleet.

[*Architecture*, X. 16: 7]

Victories such as these, Vitruvius concludes, prove that "not by machines but in opposition to the principles of machines, has the freedom of states been preserved by the cunning of architects."

Endowed with this professional credo, Roman engineers literally paved the empire with their structures: a network of finely graded highways—some still in service today—permitted Roman legions to be dispatched rapidly to Germany, Gaul, or Britain; Roman engineers tunneled the Alps and bridged the Tiber; arenas, circuses, and amphitheaters of heroic proportions provided citizens with public settings in which to indulge their most despicable vices; in the capital, Marcus Vipsanius Agrippa, friend and technical advisor to the Emperor Augustus, oversaw the building of the city's first great public baths, a new granary, and the Pantheon, with the largest free-standing dome in existence.

Suetonius, in his evaluation of the *Lives of the Twelve Caesars*, invariably includes the public works they patronized: Whereas the miscreant Tiberius was too busy consummating his "unnatural lust" to build any engineering structures of note, and the monstrous Caligula consumed the wealth of Rome in reckless extravagance, the modest and unassuming Claudius "completed public works that were great and essential rather than numerous," including fountains, a lighthouse, the harbor at Ostia, and a pair of aqueducts. [*Lives*, 220]

Most famous among the feats of Roman engineering are the nine aqueducts, stretching nearly 300 miles, that brought up to 300 million gallons of fresh water daily into the private homes and public cisterns of the capital. The arched viaducts that carried the aqueducts across valleys and ravines served Roman subjects as a pervasive symbol of imperial authority, but the system symbolizes for us the achievements of Roman engineering skill.

The triumph of Roman hydraulic engineering leads us to the final word on the subject of technology in antiquity, which may be spoken by Sextus Julius Frontinus—surveyor, roadbuilder, and until his death in 103 or 104 A.D., Imperial Commissioner of Sewers and Aqueducts. Surveying his empire of conduits, reservoirs, plumbing fixtures, and drainpipes, he paused for a moment in his labors: "With such an array of indispensable structures carrying so many waters, compare, if you will, the idle Pyramids or the useless, though famous, works of the Greeks!" [*De Aquis*, I: 16]

This is no idle boast. Indeed, it carries us as close to the ancients' view of technology as our sources are likely to bring us. Like everyone else, the Greek- and Latin-speaking peoples of the Mediterranean required food to eat and water to drink, vehicles and roads to bring these goods into their cities, a regular supply of wine as well as coins with which to pay for it, sandals for their feet and roofs over their heads, defenses against their natural and unnatural enemies, medicinal and recreational drugs, nautical and musical instruments, and all the other requisites of "civilization." As their respective empires grew in power, however, the provision of these goods fell increasingly upon the heads and hands of the slaves captured in wars of imperial conquest. We should not conclude that technics, as demeaning, became the work of slaves, but rather that technics, as the work of slaves, became demeaning.

It is an enduring accomplishment of classical philosophy that when the ideal of a free and harmonious society admitted the technical arts, it sought to contain them within human bounds, situate them within the human community, and ensure that they were governed by a system of reason and human values. But this is as far as it ventured to go. As long as the *word* was the right of the citizen and the *deed* the duty of the slave, the technical arts were linked to servitude, and *poiesis*—the "poetry of technology"—was sung only by the unfree.

3

Technology and World Culture

The view of "technological progress"
as a linear development, in which some
restless metaphysical impulse marches inexorably
westward, is inaccurate, implausible, but deeply
ingrained. In fact, Europe made a rather

late entry into the field, and much of its subsequent technological dynamism derived from its contacts with the cultures of China, India, and Islam. Two related questions disrupt the picture of a smooth ascent toward modernity: First, what explains the relative stagnation of Western technology in the period from the rise of Christianity through the end of the Middle Ages? Second, how did Europeans acquire the tools, texts, and techniques that enabled their technological revolution to finally occur? Social, political, and economic factors have been amply studied; we shall turn for clues to the world of art and ideas, and we shall begin with the texts that form the spiritual center of Christendom.

Jesus spoke to his disciples in parable and paradox, but on one point his message seems unambiguous. The Gospel of Mark foretells the final convulsions of heaven and earth and the coming of the son of man; asked when all these things are to be accomplished, Jesus solemnly declares,

> Truly I tell you that this generation will not pass before all these things are done.
>
> [Mark 13: 30]

In Matthew the Master alerts his disciples that "some of those who stand here will not taste of death" before the coming of the Kingdom of God. [Matthew 16: 28] Pressed to be more specific, he again replies,

> Truly I tell you that this generation will not pass before all these things are done.
>
> [Matthew 24: 34]

The Gospel of Luke likewise foretells "the son of man coming upon the clouds with power and great glory":

> Truly I tell you that this generation will not pass by before all these things are done.
>
> [Luke 21: 32]

And Revelations, surely the most stirring poetical achievement in all of Christian literature, opens with a terse warning: "For the time is near." [Revelations 1: 3]

To the downtrodden Greek and Roman and Jewish converts who heard them, the tenor of the Gospels was fiercely apocalyptic. Indeed, it was intrinsic to the very structure of early Christianity that redemption was imminent, for herein lay its appeal to the poor, the hungry, "you who weep now" and "you who are cast out." "You do not know the day, or the hour," adherents were cautioned, but rest assured that the Kingdom of God is at hand.

A theology of apocalypse entails considerable risk, however, for if "this generation will not pass" before its promises are redeemed, it stands or falls on its capacity to deliver the goods. This posed a major problem for the nascent church: A generation after the fact, the powers of the sky still had not been shaken; and another generation passed, and then another and another. Two rescue operations were mounted in those early centuries to hold together the community of the faithful until this awkward situation could be resolved. One, which found its most powerful expression in the writings of St. Augustine, resigned itself to the continued existence of the temporal world but urged Christians to dwell spiritually within the heavenly City of God. The other prescription called forth by the persistence of the earthly city was to hunker down and wait; this was the tactic of St. Benedict and the monastic movement to which he gave his energies and his name.

Augustine, Bishop of Hippo in North Africa and the greatest of the early Christian Apologists, composed his massive *City of God* in the aftermath of the sack of Rome in order to secure the foundations of Christian faith against the turbulent world around him. "Life in the earthly city is that of a captive and an alien," he wrote, and though our bodies must dwell here, our spirits should repose rather in the Heavenly City of the faithful. [*City of God*, Bk. 19, ch. 17] To be sure, the "city of this world" is not a wholly evil place, for God has infused the capacity for reason into even the pagan and the heretic and endowed human nature with "the power of inventing, learning, and applying . . . innumerable arts and skills which minister not only to the necessities of life but also to human enjoyment." [City of God, Bk. 22, ch. 24] Augustine even produces an inventory of "astonishing achievements" that testify to the excellent technical capacities with which God has endowed the human mind: cloth-making, architecture, agriculture, and navigation; contrivances devised for the capturing and taming of wild animals; drugs discovered by medical science; even the poisons, weapons, and other infernal equipment used in wars attest to our inventiveness.

But this most uncharacteristic technological digression (tellingly reserved to the 22nd and last chapter of Augustine's immense treatise) is anything but a celebration of the technical arts, for ultimately they are no more than testimony to the grace of God. Even less is it an exhortation for human beings to apply their divine endowment of reason and intellect to their fullest limits; to do so would imply an undue preoccupation with the affairs of the earthly city. The essential theme of his work, which lays the foundation for all subsequent Christian theology, is to remind us that in the end we are "mere sojourners in this world," "wayfarers on this earth." [City of God, Bk. 1, ch. 9; Bk. 19, ch. 17] Economic productivity, with or without the application of technology, could have little relevance to transient beings such as ourselves, and it found little encouragement in the teachings of the greatest of the early Latin fathers.

During the period of Late Antiquity—an "age of anxiety," as it has been called—there emerged a sort of countertendency to the other-worldly longings of Augustine. The 6th century saw the consolidation of spiritual communities that were very much part of this world, and whose proliferation would be of the most far-reaching importance in the economic and technological history of the West. The profoundly un-Augustinian work ethic that pervaded the early monasteries is nowhere more vividly depicted than in the monastic *Rule* composed by St. Benedict of Nursia (480–567) for the orderly governance of his community at Monte Cassino. At its heart lay the precept that "idleness is an enemy of the soul," and the injunction that followed from it: "*Laborare est orare*"—"To work is to pray." [*Rule*, ch. 48]

And work they did. Reversing almost single-handedly a millennial disdain for manual labor, the aristocratic Benedict ordered his monks out into the fields and the workshops with the warning that "the Lord looks for His workmen among the masses of men." [*Rule*, Prologue] As the monks toiled in the vineyards of the Lord, the abbot, responsible both for their physical maintenance and their spiritual well-being, evolved into a combination of shop foreman, drill sergeant, and industrial psychologist. His methodical instructions provided "the tools of our spiritual craft," and the secluded monastery itself was "the workshop in which we must diligently perform all these things." [*Rule*, ch. 4] Outside its walls barbarians—Christian and pagan—ranged freely over the domains of the shattered Roman Empire; inside, where disciplined labor was the prime expression of piety, productivity soared.

The teachings of the early monastic orders are, frankly, ambiguous with respect to technology: "They are truly monks when they must live by manual labor," Benedict wrote; but if the emphasis is upon labor *per se*, then the application of labor-saving technologies is no obvious virtue. Furthermore, since no social distinctions were permitted to divide the communal society of brothers, the Abbot saw to it that "if anyone becomes proud of his skill and the profit he brings the community, he should be taken from his craft and work at ordinary labor." [*Rule*, ch. 57]

There is a paradox here, however: Devotion to industry and agriculture was intended not only to deny the devil access to an idle mind and body, but to ensure that the brothers would not be distracted by material privation. But how much is enough? and would it not be prudent for the abbot of an isolated monastery to safeguard his flock against lean times by ensuring it a modest surplus? and in seasons of prosperity why should not the fruits of their labors be applied to charitable works? or acquiring a few holy codices for the monastery's *scriptorium*? or commissioning from a local artisan a set of stained-glass windows? or remodeling the refectory? Gradually, in their drive toward self-sufficiency, the monasteries became centers of agricultural efficiency, technological innovation, and wealth.

As the spiritual archipelago of Benedictine, and, later, Cistercian and Franciscan monasteries spread across Europe, so too did the arts and industries they encouraged. It would be risky to suppose that the monks had fomented a "medieval industrial revolution," for spiritual considerations ensured that the application of technology was not, as in the 19th century, linked to the *replacement* of human labor power so much as to its *augmentation*. The astonishing technological innovations of the medieval period do, however, correspond to a radical revaluation of the dignity and the sanctity of labor. And to an extent unimaginable in previous centuries, they were sanctified by the leading theologians of the day.

No less than Hugh of St. Victor (1096–1141), master of the great abbey school of St. Victor and one of the most influential of the medieval Parisian intelligentsia, was among the first to rescue the mechanical arts from oblivion and to secure them within the framework of a systematic theology. In his *Didascalicon*, an encyclopedic treatise on the organization of the arts and of the disciplines that correspond to them, Hugh anticipates the orgies of categorization so characteristic of later Scholastic philosophy. Knowledge, he instructs, corresponds to three "manners of things": (1) the work of God, which is "to create that which was not"; (2) the work of nature, which seeks "to bring forth into actuality that which lay hidden"; and (3) the mechanical work of the human artificer, whose task is "to put together things disjoined or to disjoin those put together." This is not a simple, Aristotelian hierarchy, however; for Hugh, these three ways of knowing are but reflections of one another within a single, spiritually indivisible totality. God has ordained the works of nature, and it is to nature that the weaver, the carver, and the founder turn "when they come to devise for themselves by their own reasoning those things naturally given to all other animals." [*Didascalicon*, I: 9]

Hugh slashed through centuries of prejudice against technology, but he wielded a double-edged sword. On the one hand, he argues, the mechanical arts function to restore the lost union of the temporal with the divine. Their other function is to compensate for the frailties of the human body by copying the forms of nature, but in this respect the mechanical arts can be no more than imitative. "Man alone is brought forth naked and unarmed," Hugh observes, and "his reason shines forth much more brilliantly in inventing these things than ever it would have if man had naturally possessed them." But the technical trades—he cites fabric-making, armament, commerce, agriculture, hunting, medicine, and theatrics—cannot themselves be sources of originality or creativity.

This prejudice became institutionalized when the old cathedral schools began to broaden their activities from theology to encompass the entire *studium generale* of the medieval university. In Paris and elsewhere, the curriculum became divided between the *trivium*, comprised of grammar, logic, and rhetoric; and the advanced *quadrivium*, which included the mathematical disciplines of geometry, astronomy, arithmetic, and music. Hugh, like other medieval Schoolmen, stopped short of including the "adulterate" mechanical sciences among these seven liberal arts, but he went further than any of his contemporaries in acknowledging their dignity and importance. [*Didascalicon*, II: 20]

Hugh's daring elevation of "the movement of artifacting fire" came at a propitious moment. Had the artistic and artisinal trades not earned this degree of theological legitimacy, they could never have been mobilized in the grandest campaign of the 12th century. The austere reforming Cistercian, Bernard of Clairvaux, had for some while been inveighing against the leaders of the Romanesque churches, whose opulence tended to divert the worshipper from the Creation of God to the creations of mortal artists and artisans on earth. His message was heard across Paris by the influential Abbot Suger (1081–1151), a monk of humble birth but extravagant tastes, whose spirit was inflamed by the desire to rebuild his church at Saint-Denis. The alliance struck by these two powerful clerics marked a turning point in the history of architecture, and called into being a whole new class of technical professionals.

Suger's compromise was to conceive a church whose aesthetic characteristics would not be painted on but designed in. Around 1140 he began the rebuilding of Saint-Denis, the cramped and dilapidated resting-place of the hereditary kings of France, in accord with this radical theological and technological vision: Structure itself would take the place of the applied ornament that the ascetic Bernard had found so objectionable, and the decorative painter would likewise yield to the engineer-architect. With a technical daring that is scarcely imaginable, the massive supporting walls characteristic of the Romanesque basilica were reduced to the barest skeleton, with external buttressing to prevent them from collapsing outward under the pressure of the great vaulted ceiling; the soaring interior space thus achieved was flooded with light transmitted through spectacular stained glass windows opened up in what had previously been thick structural walls; the presence of the divine radiated through a delicate rose window set into the west façade. On the second Sunday in July, 1144, in the presence of the King of France and his consort Eleanor of Aquitaine, the great landed nobility, five archbishops and the ecclesiastical peers of the realm, the newly-rebuilt cathedral of Saint-Denis was consecrated. A nail of the True Cross, the Crown of the Lord, and the arm of St. Simeon were deposited in a reliquary dug into the church's foundation, and the architectural vision known as "Gothic" was born.

Two principles underscored the spirit of Gothic architecture, and together they effected an unprecedented synthesis of philosophy, technology, and art. First, Suger required that all of the dimensions of the structure be equalized "by means of geometrical and arithmetical instruments"; a sacred geometry, in other words, ensured that the microcosm of the cathedral was

Figure 3.1

Remember that if you wish to build great buttress towers, they must project sufficiently. Take pains with your work, and you will act prudently and wisely.

—from Villard de Honnecourt, *Sketchbook*

Rheims Cathedral
Schematic view of the double row of flying buttresses

as structurally harmonious as the macrocosm of the universe. Second, on the mystical premise that God is manifest in light, the interior was made to "shine with the wonderful and uninterrupted light of most-sacred windows." [*De Consecratione*, IV] These two essentially spiritual objectives, which had the effect of "transferring that which is material to that which is immaterial," [*De Administratione*, XXXIII] could be achieved only through a practical mastery of structural principles. To be sure, the builders of Saint-Denis, and of the eighty-odd cathedrals that followed, submitted humbly to the inexorable laws of God; but they also submitted, in the words of their 19th-century admirer Eugène Viollet-le-Duc, to "the inexorable laws of statics."

The status of the technical arts had by the 12th century risen to such an extent that Suger, one of the two or three most powerful men in France, could describe with evident pride how he himself supervised the work of his craftsmen, surveyed local stone quarries, and personally ventured out into the surrounding forests in search of timber of the proper dimensions. The fabulous stained-glass windows that are the enduring symbol of Gothic architecture offer luminous testimony to the role of the technical trades in the spiritual revival of the 12th century, for they depict the craft guilds that donated them as routinely as they show scenes of the Passion: stonecutters and masons, glaziers and metalsmiths, carpenters and woodcarvers. Although scarcely anything survives to tell us how production was actually organized, the *Sketchbook* of the engineer-architect Villard de Honnecourt (about 1235) gives us some intriguing insights, as does an extraordinary treatise *On Divers Arts* by a 12th-century Benedictine who introduces himself only as "Theophilus, a humble priest, servant of the servants of God, unworthy of the name and profession of monk." [*Div. Art.*, Prologue]

Theophilus provides the first detailed description, by a practicing artisan, of the technical crafts that must be mastered by whoever would undertake to embellish the house of the Lord: For the painter he details the composition of pigments, inks, glues, resins, and enamels; he describes the furnaces, molds, and techniques of the glassmaker; and with saintly patience he enumerates the 42-odd metalworking tools needed to cast a good bell. Specifications are always qualitative for Theophilus, since standardized units of length, weight, and time did not yet exist. Thus nails might be "a finger long" and a vent "big enough to put your hand through," and some rather unusual practices are indicated: The preferred technique for hardening an engraving tool is by plunging it, red-hot, into the urine of a small red-headed boy. [*Div. Art.*, III: 21] The technology of Theophilus is not that of a theorist but of the master craftsman, working by the rule of his thumb and the seat of his pants, and in full confidence that "whatever in the arts you can learn, understand, or devise, is bestowed on you by the grace of the seven-fold Spirit." [*Div. Art.*, III: Prologue]

This is a most significant bit of editorializing, for it suggests that as late as the 12th century Theophilus numbered among his pious readers not a few who might be suspicious of his worldly zeal for "the practice of the divers useful arts." For spiritual ammunition he turned to the Holy Scripture and the injunction (Ecclesiastes 1: 18) that "He that increaseth knowledge, increaseth labor." But there is a rhetorical sleight-of-hand here: The meaning of the Latin *laborem* is not so much "productive labor" as "burdensome toil." The Bible, we will recall, suggests a strong connection between knowledge, sin, and suffering—an equation that our industrious monk has neatly reversed. Clearly, no Christian technology was possible without Christian theology.

This was not, however, the wholly detrimental relationship we moderns might assume it to have been. To be sure, when Christianity embraced an "other-worldly asceticism" it achieved profound spiritual insights but turned the community of the faithful away from the mundane world of necessity, which, as Hugh of Saint Victor reminds us, "hath mothered all the arts." [*Didascalicon*, II: 9] But when it embraced human life in its manifold complexity it tended to endorse the direct interaction with nature—through labor, experiment, and invention—that is of the essence of technology.

The Franciscan friar Roger Bacon (c. 1214–94) took the elements of the medieval synthesis of reason and faith, science and spirituality, theology and technology, as far as they could go before they began to run afoul of each other. In his encyclopedic *Opus Majus*, he attempted to convince the Pope himself of the value of the methods he had been teaching at the recently founded university at Oxford. The new experimental science, Bacon wrote to Clement IV in 1268, tests the conclusions of the speculative sciences by means of experiment; it thus provides the ultimate key with which to unlock the secrets of nature, and has, moreover, the capacity to provide Church and state alike with "instruments of wonderful utility." Bacon envisioned a science that would place at the disposal of the state biological weapons that "act by means of an infection" and terrible explosives, derived "from the force of the salt called saltpeter," which produce so horrible a sound that no enemy will be able to resist. The Church, for its part, would benefit from the new experimental methods, for by them the enemies of the faith might be destroyed "rather by the discoveries of science than by the warlike arms of combatants." [*Opus Majus*, vol. II, VI: p. 633] For individuals, he promised, the new science will provide perpetual baths, ever-burning lamps, ships that can be guided by one man, a device for flying, bridges without columns or supports, and a car that will move at "inestimable speed" without the help of any living creature. "These devices are certain," he assured his would-be patron; "I am acquainted with them explicitly, except with the instrument for flying which I have not seen. But I know a wise man who has thought out the artifice." [*Epistola*, p. 27]

The visionary figure of Roger Bacon stands at a crossroads in the history of technology. He could celebrate the practical application of reason to nature, as he believed that "the end of all true philosophy is to arrive at a knowledge of the Creator through knowledge of the created world." But technology placed in the service of theology proved to be a risky business. Already in his lifetime the Bishop of Paris had moved to condemn 219 heterodox doctrines, and Friar Bacon himself would languish in prison for fifteen years, charged with the very practice of magic whose "nullity" he had so passionately argued.

The high Middle Ages were marked, nonetheless, by an extraordinary and unprecedented burst of technological dynamism. Technical innovations had already revolutionized agriculture, warfare, energy production, and transport, and even more far-reaching changes would take place before the allied forces of war and pestilence arrested the wave of technical progress in the 14th century. At the same time, much of the accomplishment of the Christian Middle Ages owed to the ongoing exchange of travelers, merchants, pilgrims, students, missionaries, soldiers, craftsmen, diplomats, and adventurers whose commerce knitted together the rival cultures of Judaism, Christianity, and Islam. "Technology transfer" and the "diffusion of innovation" are modern terms, but they describe a process as old as technology itself. Moreover, the direction of traffic was characteristically from East to West.

While monastic copyists, working without the benefit of eyeglasses or paper, labored to preserve fragments of antiquity, a messianic movement was taking shape in the commercial cities of Arabia that would prove decisive for the growth of technology in the West. One hundred years after the death of Muhammad in 632 the advance of militant Islam was checked at Poitiers by Charles Martel, but Arab armies had already carried the teachings of the Prophet to a community of the faithful that stretched from the Indus River to the Atlantic Ocean. Five hundred years later, by the time the Abbot Suger blessed Louis VII from his restored church at Saint-Denis and sent him off on the Second Crusade, the Muslim had become not just an infidel but the antithesis of every Christian virtue:

Knightly epics such as the *Song of Roland* parade the peoples of Islam as idolotrous worshippers of the unholy trinity of Muhammad, Apollo, and a mythical Tervagant, but this is nothing more than a grotesque caricature rooted in fear and ignorance. The first translation of the *Koran* was commissioned in 1143 by

Peter the Venerable, Abbot of Cluny, who required a better knowledge of the heresies with which he was locked in mortal combat. From it Christian scholars might have learned that Islam teaches that "the ink of the scholar is worth more than the blood of martyrs," and that Muhammad was only the last in a prophetic succession that also included Abraham, Moses, and Jesus. By that time, however, the Muslim had became cast as the Christian's mortal enemy:

> Evil at heart,   and guilty of great crimes,
> He has no faith   in Mary's holy Son.
> This pagan's skin   is black as melted pitch.
>
> [*Roland*, CXIII: 1472–4]

Only in the sphere of science and technology did Europe, Byzantium, and Islam form a single cultural unit, and in these fields alone were Europeans able to cultivate an objective attitude toward Islam.

The Arab world made two indispensable contributions to the growth of science and technology in the west. First was the recovery of Greek natural philosophy, the greater portion of which reached the Latin-speaking west not directly but through re-translation from the Arabic. Beginning in 9th-century Baghdad, Arab translation centers spread from southern India to southern Spain, and by the end of the 12th century, European scholars were again studying the scientific texts of Aristotle, Hippocrates, Euclid, Galen, and Ptolemy, whose *Mathematical Treatise* remains known by its Arabic title, *al-Majisti* (Latinized to *Almagest*).

Second, their drive to regain access to the ancient authorities inevitably brought medieval scholars into contact with works of the Arab guardians of that tradition, the Muslim *Faylasûfs* (philosophers) known to the Latin west as Alpharabius (al-Fârâbî, 870–950), Avicenna (Ibn Sînâ, 980–1037) and Averroës (Ibn Rushd, 1126–98). In Spain, Sicily, and across North Africa, Jewish scholars contributed significantly to the advance of science and learning in their capacities as both originators and intermediaries. Moses ben Maimonides (Ibn Maymûn, 1135–1204), one of the most respected scholars of the entire medieval period, led the attack against astrology and other pseudosciences, and published distinguished treatises on medicine—although even this great rationalist was not above adding a dash of pulverized mouse brains to his list of cures.

From the standpoint of technology, the Golden Age of Islam may be said to have begun with the brothers Musâ, who flourished at the 9th century court of the Caliph of Baghdad. Sons of the noted astronomer and highwayman Musâ bin Shakir, the three brothers—Muhammad, Ahmad, and al-Hasan—were among the first resident scholars at the famous *Bayt al-hikma*, the "House of Wisdom" founded by the Caliph al-Ma'mûn after Aristotle appeared to him in a dream and assured him that the exercise of reason is not prejudicial to the Good but the surest route to it. On the basis of this revelation the Caliph and his successors committed themselves to the institutional support of research, experiment, and translation—a program which the Banu ("sons of") Musâ did more than any others to advance.

When they were not experimenting with war engines, astrolabes, and musical automata, the tireless Banu Musâ brothers managed large-scale engineering projects, arranged stipends for promising younger scholars and pensions for their elders, confiscated the libraries of their rivals, and sent expeditions to Byzantium and beyond in search of Greek scientific manuscripts. They also provided the earliest textual evidence of the Islamic fascination with the machine: *The Book of Ingenious Devices*, composed and painstakingly illustrated by Ahmad Musâ with the assistance of his two brothers sometime in the mid-9th century. Given the aridity of the region, it is not surprising that their central preoccupation was with devices controlling the circulation of water: complicated fountains; self-regulating troughs, basins, and vessels; boilers; a breathing apparatus that will allow a person to descend safely into a contaminated well; a dredging machine by which "things that fall into wells or are submerged in rivers, seas, and so on may be extracted." [*Hiyal*, model 100]

The contrivances of the Banu Musâ were actuated by rack-and-pinion gears, horizontal water wheels, conical valves, concentric siphons, and assorted floats, pulleys, and cranks that had not yet entered the technological vocabulary of European engineers. Their designs, however, include only a few public works and are taken up rather with toy automata, trick vessels, and amusing mechanical contraptions with which, if God wills, "the Adept" can deceive and delight a courtly audience.

This apparently conservative stance toward the application of sophisticated mechanical knowledge to productive work may have deep cultural roots. As in the Europe of monks and abbots, medieval Islam had to ensure that the mechanical sciences were fitted into a larger framework of values—in the words of a 10th-century Muslim cleric, that "there is no contradiction at all between them and the religious sciences." [al-'Amiri, pp. 263–4] In the theocratic order of Islam, where the boundary between sacred and profane had been systematically eradicated, an uncontrolled pace of technical innovation may have been seen as disruptive to the higher values of stability, tranquility, and peace.

The delight in problem-solving and the spirit of intellectual play can be found at the other end of the Islamic Enlightenment in the important but shadowy 12th-century figure of al-Jazârî, who worked as a "court engineer" specializing in the construction of water-raising machines, fountains, clocks, technical instruments, combination locks, and mechanical toys. Al-Jazârî relates that he was summoned by his royal patron—al-Sâlih Nâsir al-Dîn Abî al-Fath Mahmûd bin Muhammad bin Qarâ Arslan bin Dawûd ibn Sukmân bin Artuk, King of Diyâr Bakr in modern Turkey—with the following instructions: "You have made peerless devices, and through strength have brought them forth as works; so do not lose what you have wearied yourself with and have plainly constructed. I wish you to compose a book for me which assembles what you have created." [Kitab, p. 15] The result was al-Jazârî's encyclopedic Book of Knowledge of Ingenious Mechanical Devices, prepared in about 1206. It survives as a compendium of the mechanical apparatus available to a

12th-century Muslim and a testament to four centuries of official indulgence of technological artistry and play.

Scholars have long since disposed of the notion that the growth of modern technology followed some neat European trajectory, much less that it owes to some peculiarly European genius. It is enough to list the familiar terms that have entered our technical vocabulary from the Arabic—"algorithm" from the name of the 12th-century mathematician al-Khuwârizmî and "algebra" from the title of his theoretical treatise, *al-Jabr*, followed by a host of terms used in chemistry (alcohol, alkali), astronomy (azimuth, zenith), and navigation (admiral, arsenal), as well as in agriculture, textile manufacturing, and commerce (the modern "check" derives from the Arabic *sakk*). Only rarely, however, are we able to trace the precise route by which ideas or artifacts of technological significance were transmitted across cultural boundaries: In 1201 the Pisan merchant Leonardo Fibonacci composed a treatise in which he expounded the Indian ("Arabic") system of numerals—including the all-important use of "nothing" (zero, *sifr*) as a place-holder—which he had mastered during a period of residence in North Africa, but this is a rare exception. For precisely this reason the few first-hand accounts we possess are of inestimable importance.

Foremost among these is surely the memoir of Marco Polo, who set out from Venice with his father and uncle in 1271 and returned in 1295 after a 24-year sojourn at the court of the Kublai Khan. Polo's account reveals to us, more vividly than any contemporary source, the technological fecundity of the east Asian civilizations and offers the most suggestive hints as to the routes by which technology diffused westward. He appears personally to have brought back to Europe knowledge of "the black stones that are dug in Cathay and are burnt for fuel" (coal), of the substance of a vein found under the mountains of Chingintalas "that when crushed into a fiber is resistant to fire" (asbestos), and of the notes bearing the seal of the Great Khan with which his subjects may transact their business "just as if they were coins of pure gold" (paper money). [*Travels*, II: XXX; I: XLII;

Figure 3.2

It is a machine for raising water from a pool or a well by an animal who rotates it.

Al-Jazârî, *Book of Knowledge of Ingenious Mechanical Devices* (c. 1206)

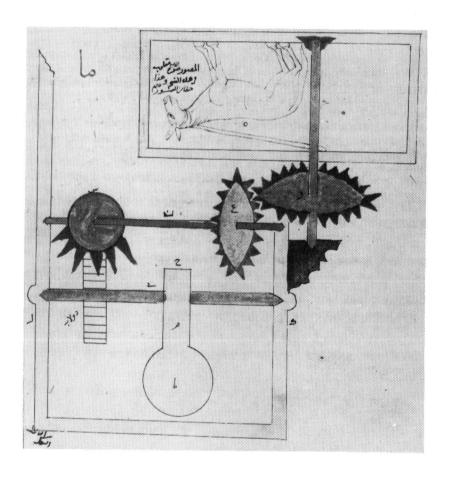

The caption reads: "The artist has drawn the animal upside down in this Section. This is a mistake—understand."

II: XXIV] Even this native Venetian was awed by the sophisticated bridges and canals that graced the imperial capital.

The cultural and linguistic barrier that separated the European and Arab-speaking peoples from the ancient civilizations of East Asia delayed the transmission of scientific concepts, but did not to the same extent impede the transfer of technologies. We now know that not only gunpowder, the elements of printing, and the magnetic compass—the three inventions of which Francis Bacon was so proud—originated in China, but also the wheelbarrow, cast iron, the suspension bridge, mechanical clockwork, and countless others. Arab traders, Portuguese Jesuits, and Mongol horsemen were among the carriers of Chinese technologies across the old silk routes of Asia to the West, where in the more fluid social environment of Europe their impact proved revolutionary.

The contributions of Islam and China to the world's store of science and technology are a matter of historical record, but the brilliance of their respective accomplishments must not obscure the fact that the breakthrough to modern, science-based technics occurred in Europe, and in Europe alone. Neither in China nor in the Islamic world did the traditions of investigation and invention usher in the mechanistic, mathematicized view of nature characteristic of modern science, nor did they issue in an industrial revolution. The reasons for this are formidably complex and must be sought in the peculiarities of social, economic, and even natural history. But since our concern here is with the realm of ideas, it may be useful to point to some features of the non-European intellectual environments that seem especially relevant to science and technology.

Chinese civilization was for two millennia dominated by the official Confucian school, whose ideal is embodied in the figure of the scholar-gentleman dedicated to a life of ethical cultivation. Paradoxically, however, while "democratizing" the route to power and influence, the Confucian ethos of enlightened public service may have had the effect of channelling the most intellectually gifted into traditional China's vast administrative bureaucracy and away from the commercial and technical professions which proved to be the source of technological dynamism in Europe.

Whereas Confucianism tended to seek for human excellence within the social order, the countertradition of Chinese intellectual history, Taoism, affirmed the order of nature, unified and independent of human standards. This attitude is in no way inconsistent with the requirements of a scientific world view, but the mystical and contemplative character it assumed in practice is illustrated by a parable from one of the earliest Taoist texts. Asked why he does not take advantage of a simple labor-saving machine, a farmer laughs and replies,

> I have heard from my master that those who have cunning devices use cunning in their affairs, and that those who use cunning in their affairs have cunning hearts. Such cunning means the loss of pure simplicity. Such a loss leads to restlessness of the spirit, and with such men the Tao will not dwell.
>
> [*Chuang Tzu*]

We must marvel that in the face of such suspicion the technological achievement of traditional China was as impressive as it now appears to have been.

If the trajectory of medieval technology so characteristically has Chinese origins, it frequently reached the west through Arab transmission. Here again, however, we find that despite a formidable legacy, by the 12th century the pace of technological dynamism had slowed dramatically in the Islamic lands relative to Europe. This need not imply that once it passed the baton westward Islam had fulfilled its function and could retire from the scene of world history; to the contrary, Muslim scholars have argued that the goals of stability and equilibrium have been mistaken by Westerners for stagnation. Whatever may be its resolution, the roots of this controversy can be traced back to the greatest intellectual duel of medieval Islam, which pitted the conservative theologian al-Ghazâlî against the rationalist-

Aristotelian Averroës, and which corresponded to a decisive turning point in the history of technology.

The first salvo in this debate was fired by al-Ghazâlî (1058–1111), who launched a frontal attack against the rationalist current that had been visible in Islam since its earliest encounters with Greek philosophy. In *The Incoherence of the Philosophers* and other tracts, the preeminent figure in the history of Islamic orthodoxy argued against the pretensions of philosophy to yield secure knowledge about nature, much less to provide the tools with which to control it: "Sun and moon, stars and elements are in subjection to [God's] command," and not to natural law or human design. [*Deliverance*, p. 37] From Córdoba he was answered by Averroës, who countered that the rigorous investigation of nature was in no way incompatible with revealed truth. [*Harmony*, ch. 2] Averroës' attempt to reconcile natural reason with religious faith was rejected by the Islamic world, although it would have a powerful impact upon Christian thinkers such as Thomas Aquinas. Since that time Muslim orthodoxy has rejected the purely quantitative, rationalistic, and materialist picture of nature and the belief in open-ended technological progress that corresponds to it.

The scientific and philosophical studies championed by the embattled figure of Averroës spread gradually northward, where they catalyzed the scientific and technical learning of the newly founded universities of Bologna, Paris, and Oxford. Indeed, Europe proved to be the ultimate beneficiary of the scientific legacy of Islam. A concrete explanation for the leveling-off of scientific and technological activity in the Islamic world was offered by the 14th-century Muslim thinker Ibn Khaldûn, whose meditations on the rise and decline of civilizations have established him as the founder of the philosophy of history. How could it have come to pass, he asked in the afterglow of the Islamic Enlightenment, that "the Arabs, of all people, are least familiar with crafts?" His answer, though rich in philosophical subtlety and historical detail, can be briefly summarized. Technology thrives only where there exists a large and sedentary civilization, whose cultural traditions are deeply rooted and in which technical knowledge can be transmitted over many generations. In contrast to the peoples of China, India, and Christian Europe, "the Arabs are more firmly rooted in desert life and more remote from sedentary civilization, the crafts, and the other things which sedentary civilization calls for." [*Muqaddimah*, I. 5: 20] Although Ibn Khaldûn is known for his strikingly modern, wholly materialist interpretation of Islamic history, the victory of the orthodox theologians cannot be discounted.

The testimony of the medieval period reveals that technology is disrespectful of cultural boundaries and resists the claims of national pride. But cultures are social as well as geographical entities, and generalizations about "technology and world culture" are necessarily partial. This was the message of the late medieval writer Christine de Pizan, whose *Book of the City of Ladies* was composed in 1405, and who offers a dissenting perspective on medieval ideas and engineering.

The literature of her day, she found, perpetrated an ongoing attack against women, "the vessel as well as the refuge and abode of every evil and vice." The effect, if not the intention of this slander, was to remove women from the great spiritual and practical achievements of civilization. While pondering this gloomy situation, Christine was visited by the crowned figures of Reason, Rectitude, and Justice, who instructed her to build an allegorical "Cyte of Ladyes" as a tribute to the role of women in the building of the human city.

This proves to be a technological no less than a political and cultural edifice. With "the pick of understanding" and "the trowel of your pen" Christine begins to construct a city peopled with women of political power, military valor, poetic vision, and moral dignity. She also finds, however, that history is littered with allegories of women who have mastered the technical arts: Ceres, who taught her subjects how to cultivate the earth; Arachne, who invented the prototypical technology of weaving;

Minerva, who invented shorthand, addition, and—though "it is far removed from a woman's nature to conceive of such things"— the technique of making steel armor. [*City of Ladies*, Part I] With boldness and imagination Christine sought to retrieve the accomplishments of women from myth and restore them to a place of honor in the medieval city.

Christine de Pizan could not believe that God—the "Divine Artificer" of the industrious Middle Ages—had willfully constructed so imperfect a vessel as women were alleged to be and thus built her case for women's contribution to technology on strong theological foundations. This gesture of piety is wholly characteristic of the medieval era, in which technology made immense strides but reached a point beyond which it could not go. For Western technology to be unleashed, in all its fury and magnificence, it would first have to shake itself free from religious dogma, political ideology, and metaphysical speculation. Once the emancipation of technology was achieved, however, it became the task—our task—to restore to it some measure of human guidance.

4

■ ■ ■ ■ ▦ ▪

Head and Hand in the
Culture of the Renaissance

The Renaissance has been claimed—rather

unfairly—it seems, by historians of art and

literature, the fields that testify most brilliantly

to the "rebirth" of classical models lost

or suppressed during the Christian Middle Ages.

The technology of the era, however, was characterized less by the revival of ancient wisdom than by a series of irreversible breaks with the past. The technical achievements of antiquity—Greek and Roman but also Chinese and, somewhat later, Islamic—were indeed impressive, but their most striking characteristic was an almost total divorce between scientific explanation and practical applications. To the extent that it laid the foundations for a reconciliation between head and hand, the Renaissance marks not a rebirth but a radical departure.

The first hint of the new directions technology was to take was announced by the humanist scholar Leon Battista Alberti (1404–72), the original *uomo universale* — the "universal man" of the Renaissance—whose writings on painting, sculpture, and architecture are all that remain of a vast literary and scientific output. In his treatise on the art of building, the first systematic exposition on the subject since Vitruvius, Alberti sounded a note that stated with bold assurance the spirit of modernity. "Who is to be called an architect?" he asks rhetorically:

> I will call an architect one who, by sure and wonderful Art and Method, knows *first*, how to divide things with his mind and intelligence, *secondly*, how rightly to put together in the carrying out of the work all those materials which, by the movements of weights and the conjoining and heaping up of bodies, may serve successfully and with dignity the needs of man.
>
> [*Arch.*, Preface]

Whether he is discoursing about vaults or domes, stone bridges or staircases, Alberti's demand for the application of reason to whatever art one practices is unrelenting. There is no inclination to submit humbly to the "sacred geometry" of the medieval cathedral builders; now it is human reason itself, infinite in all directions, that is the divine endowment.

This integration of scientific theory and technological practice would become the defining feature of the West. It was not an inevitable outcome of intellectual progress, however, for Alberti's call for the application of reason to art and industry had its roots in specific social changes that marked the later Middle Ages. Most important among them was the rapid growth of the towns, which called forth a new class of technical professionals, skilled in industrial and artisanal trades and protected by the system of medieval guilds. The decline of feudal institutions in the 14th century and the concomitant rise of powerful merchant families loosened the bonds of this technical elite, who found their services in considerable demand by the newly rich and fiercely competitive princely houses. There emerged, in the mid-15th century, a wholly new social type: the artist-engineer, cast adrift and dependent now upon the patronage of princes, prelates, and patricians.

A less-than-flattering picture of these Renaissance princes, and of their technical requirements, was painted by Niccoló Machiavelli (1469–1527); although more interested in the machinations of princes than in their machines, he was insisting at just this historical moment that politics in the modern age must be guided not by moral pieties but by technical expertise. Above all, the prince who would rise above "the extraordinary and inordinate malice of fortune" must control the means of violence, because "there is simply no comparison between a man who is armed and one who is not." [*Prince*, ch. 8, 14] Moreover, although he is often misrepresented as an advocate of self-serving despotism, Machiavelli always insisted that the prince must at the same time provide for the well-being of his subjects (if only to defuse their rebelliousness). The master of Renaissance statecraft, then, must have at his disposal the services of military as well as civil engineers, and in the Italian city-states of the late 15th century such people were in plentiful supply.

There exists a letter, addressed by just such a man to the powerful Ludovico Sforza, Duke of Milan, that reflects the new world of courtly patronage. "Most Illustrious Lord," wrote this unemployed engineer to the *Condottiere* in 1482:

Having now sufficiently considered the specimens of all those who proclaim themselves skilled contrivers of instruments of war, and that the invention and operation of the said instruments are nothing different from those in common use, I shall endeavor, without prejudice to anyone else, to explain myself to your Excellency, showing your Lordship my secrets, and then offering them to your best pleasure and approbation.

The applicant then enumerated, "with the utmost humility," the technical skills he was prepared to place at the disposal of the prince, and even promised that his expertise was not limited to fortifications, water-delivery systems, and courtly spectacles: "I can carry out sculpture in marble, bronze, or clay," he boasted, "and also I can do in painting whatever may be done, as well as any other, be he who he may be."

This brash young Tuscan may be forgiven his youthful arrogance if we confess that he is none other than Leonardo da Vinci, who, weary of the intellectual pretensions of his native Florence, had sought employment in a court where learning was valued insofar as it was applied to some useful end. His own estimation of his abilities has been shared by his admirers ever since Giorgio Vasari, in his *Lives of the Artists* (1568), gushed that "Nature favored him so greatly that in whatever his brain or mind took up he displayed unrivalled harmony, vigour, vivacity, excellence, beauty, and grace." [*Lives*] Ironically, however, this unrestrained adulation has tended to obscure, rather than clarify, the dramatic turning point represented by the multifaceted personality of Leonardo.

It is true that Leonardo tried his hand at everything from costume design to irrigation systems, wrote a (lost) treatise on singing and painted the most famous portrait in the world. From the standpoint of technology, however, Leonardo was not so much a timeless genius who boldly transgressed all boundaries as a particularly talented exemplar of a type that had become common in the city-states of the Italian *Quattrocento:* the engineer-artist, versed in the sciences but always with an eye to their application in the service of his employer. Judged in modern terms, Leonardo da Vinci defies all categorization; judged by the standards of his own contemporaries, when he fused art, science, and technology into a single, seamless career he was just doing his job.

Though raised and apprenticed in the gilded Florence of the Medici, Leonardo felt out of place in its academic circles, for he was not, as is sometimes assumed, a great scholar steeped in the wisdom of the ages. His Latin was poor and his Greek non-existent, and he remarked (somewhat defensively, no doubt) that the ability to quote ancient authorities demonstrated a good memory, but not necessarily a good mind. To the contrary, he felt that the neo-Platonist scholars who were reviving the legacy of antiquity were reviving also the ancient prejudice that divorced abstract speculation from application and experience. "It seems to me," he countered, "that those sciences are vain and full of error which do not spring from experiment, the source of all certainty," and that for this reason "mechanical science is of all the most noble and the most useful." [*Notebooks*, II: 1154]

Such bold claims for the intellectual and social respectability of technology ran counter to the inherited traditions of the ages. In antiquity, the mechanical arts were the business of slaves, and in the Middle Ages, they were still considered "illiberal"—arts whose ties to material need made them unbefitting of the leisured or liberated individual. For the Renaissance engineer, however, whose finest representative was Leonardo da Vinci, *doing* was not only on a par with *knowing*—it was the ultimate way of knowing. Very rarely do we find him attempting to derive technical solutions from theoretical principles; rather, the transmission belts, escapements, and gear assemblies that fill his notebooks made manifest the physical principles according to which they operated.

Indeed, although incautious admirers have made him into a "precursor" of Galileo, Newton, and Einstein, Leonardo's interest in theoretical problems was always subordinated to his fascination with concrete problem-solving. His praise of mathematics

Figure 4.1

And if they despise me who am an inventor, how much more should they be blamed who are not inventors but trumpeters and reciters of the works of others.

Leonardo da Vinci, *Notebooks*

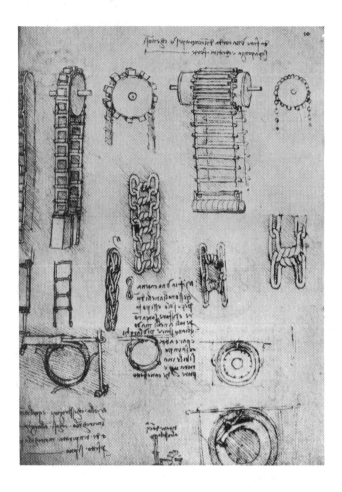

Studies for transmission chains

was never equalled by his command of it, and he seems never to have attempted a grand theoretical synthesis. "It is as if he looked at everything as an engineer rather than as a theoretical student of nature," sniffed E.J. Dijksterhuis in his classic study of the rise of the modern scientific worldview. This appraisal seems reasonable in everything except its tone of evident disappointment that so noble a thinker could wallow in such ignoble artistic and mechanical pursuits.

We are told that Leonardo rarely ventured out into the streets without his sketchbook and that nothing failed to excite his imagination. In a spirit of curiosity that seems never to have abated, he filled 5,000 pages with studies of birds in flight, eddies of water in a gutter, the musculature of laborers, the phases of the moon. "Leonardo was so delighted when he saw curious heads," marveled Vasari, "that he would follow about anyone who had thus attracted his attention for a whole day, acquiring such a clear idea of him that when he went home he would draw the head as well as if the man had been present." [*Lives*] His passion for observation required first that he learn how to see—"the eye is the window of the soul," he wrote. It may be that his principal accomplishment was not the *Mona Lisa* or the flying machine or his brilliant anatomical drawings, but his ability to see through the prejudices of the past.

Leonardo was not a lone visionary who overstepped the bounds of his age, but very much a spokesman for it. As the names of his colleagues and contemporaries have become known to us, it has become increasingly clear that he was riding a wave of invention and innovation that included the printing press, the cannon, and the mechanical clock, and all that they in turn made possible. As often as not, these engineer-artists themselves have provided the best historical evidence of the mounting European commitment to technology, for the flourishing Renaissance book trade encouraged the publication of their technical treatises: the *Pirotechnica* of Vanoccio Biringuccio (1540), which applied the principles of chemistry to industry; *De Humani Corporis Fabrica* (1543) by the Flemish anatomist Andreas Vesalius, which sought to heal the breach between the scientific study of the human body and the techniques of surgery and dissection; *De Re Metallica* (1556), in which the German Georgius Agricola attempted likewise to bring scientific methods to the technologies of mining and metallurgy; and, continuing the Italian tradition, *The Various and Ingenious Machines* of Agostino Ramelli (1588), a virtual encyclopedia of Renaissance mechanics. Invariably, the sinews and ligaments, epicyclic gear trains, and experimental apparatus illustrated in these classics in the history of technology are set against a human backdrop—pastoral landscapes littered with villages, antique ruins, and industrious townsfolk—as if to emphasize that all engineering is in the last analysis *civil* engineering, the engineering of civilization.

The determination of these pioneers to find a scientific basis for their empirical techniques was less than fully realized. Alberti scoffed at the superstitions of ancient builders, but conceded that in laying foundations, the counsel of astrologers "may be of great service if true, and can do little harm if false"; [*Arch.*, II: 13] Agricola described the technical procedures for expelling pestilential vapors from mineshafts, but advised that "demons of ferocious aspect" can be put to flight only by fasting and prayer; [*Metallica*, VI: p. 217] and the precocious Pico della Mirandola (1463–94), who sought to demolish all traces of medieval prejudice and obscurantism, still recognized two forms of magic: "One consists wholly in the operations and powers of demons, and this consequently appears to me, as God is my witness, to be an execrable and monstrous thing. The other proves, when thoroughly investigated, to be nothing else but the highest realization of natural philosophy." [*Oration*, p. 53]

The case of Pico, the young Count of Mirandola, is particularly suggestive of the transition to a modern science-based technology whose foundations were laid in the Renaissance. In 1486, in an act of intellectual bravado that has never been equalled, the 24-year-old nobleman drew up 900 theses on all aspects of philosophy, science, and law and challenged leading thinkers from all over Europe to come to Rome to dispute them. Pico exemplifies the spirit of Renaissance "syncretism," the belief that there are no conflicts that will not yield to philosophical resolution and that all cultures, ages, and schools of thought have some wisdom to contribute in this conciliatory task: mathematics, but also numerology; physics, but also metaphysics; astronomy, but also astrology; experimental chemistry, but also alchemy and magic.

The occult sciences that flourished in the Renaissance occupy an important place in the history of technology. Their aim, like that of all technology, was the willful manipulation of nature (even though it may have been through the Philosopher's Stone or the Elixir of Life or an appeal to the World Soul). Their collective legacy is ambivalent, however. The alchemist's laboratory contributed a variety of experimental apparatus, materials, and procedures; likewise, the mystical belief in perfect harmonies and "correspondences" between macrocosm and microcosm may have gradually enhanced the sensitivity to mathematical laws of nature that was to mature in subsequent centuries. The most enduring legacy of the alchemical tradition, perhaps, lay in the claim of Paracelsus (1493–1541), most famous of the Renaissance techno-magicians, that "theory and practice should together form one," which helped implant the ethic, so foreign to the ancients, that speculative natural philosophy may be put at the service of humane practical ends. [*Paracelsus*, 51]

On the other side of the ledger, the obvious theoretical inadequacy of the mystical sciences placed a fixed, upward limit on their possible achievements. The thinking of their adepts tended to be analogical rather than analytical, their veneration of ancient authorities kept them mired in a world of four Aristotelian elements, Hippocratic bodily humors, and the mystical *Corpus Hermeticum* of the legendary Hermes Trismegistus. And the same superstitions that underlay both the practice and the persecution of witchcraft permitted the outright murder of perhaps 300,000 women accused of casting spells before the frenzy died down around 1700.

In the meantime, however, the figure of the "magus" preoccupied the life and letters of the Renaissance; indeed, he entered the mainstream of European literature less through his own arcane utterances than through the writings of his admirers and critics. The most vibrant tradition swirled around the alleged exploits of a German mountebank named Faust, first chronicled in 1587, some fifty years after his presumed death.

In fiction, drama, and music, the theme of the Faustian Bargain has served as an outstanding index of the changing perceptions of technology, and of its long and dubiously fruitful implication with sorcery, magic, and witchcraft. In *The Historie of the damnable life and deserved death of Doctor John Faustus* the story is first told of an unprincipled conjurer who dabbled in the black arts until he acquired sufficient skill to summon up the devil himself and to bargain with him. A clear message pervades the earliest Faust legends: Infatuated with his technical virtuosity, he had transgressed the bounds of morality and religion.

When the diabolical exploits of the great necromancer were taken up by the Elizabethan dramatist Christopher Marlowe (1564–1593), the story entered the mainstream of European literature. Marlowe's interpretation of *The Tragicall History of the Life and Death of Doctor Favstvs* was first performed around 1588, a few years before he was stabbed to death in a tavern brawl.

Marlowe's Faustus, though a learned Doctor of Philosophy (and of Law, Medicine, and Theology), has grown weary of the sterile science preached and practiced in the university, and rightly so: disdaining the experimental approach and fettered by the orthodoxies of Aristotle and the Church, the medieval doctors had reached limits beyond which they could not go. Thus Faustus takes his stand: "Divinity adieu," he cries, and burning his bridges behind him, turns to the sciences of the occult in the service of his ambitions:

> O what a world of profit and delight,
> Of power, of honor, of omnipotence,
> Is promised to the studious artisan!

All things that move between the quiet poles
Shall be at my command. Emperors and kings
Are but obeyed in their several provinces,
Nor can they raise the wind or rend the clouds;
But his dominion that exceeds in this
Stretcheth as far as doth the mind of man.
A sound magician is a mighty god.

[*Faustus*, I: 48–63]

No longer content to master a few conjurer's tricks, Faustus is possessed by a vision of technological omnipotence. He will bridge the continents and subdue the elements, and become "a mighty god." But to achieve this superhuman power Faustus must compromise a measure of his humanity. At the end of his allotted 24 years he laments the curse he has brought upon himself, but his repentance comes too late: Lucifer returns to claim his immortal soul.

Only a few decades had passed since the legend of Dr. Faustus first began to circulate, but already an immense transvaluation had taken place. The earliest version of the story cast him as a wretched trickster who consorts with witches and flies through the air on a wine cask: his life is "damned" and his death "deserved," and both set an example to Christians to contain their ambitions within a finite horizon. In the hands of Marlowe, however, the fate of the sorcerer has grown to "tragicall" proportions and is clothed in a certain grandeur. He has gambled and lost, but only because his Edenic quest for knowledge could not be contained within the cloistered medieval world view.

The career of the magician is given a lighter tone in *The Alchemist* by Ben Jonson (1572–1637). Where Marlowe's tragic Faust had dreamed of a kiss from Helen of Troy ("the face that launched a thousand ships, and burnt the topless towers of Ilium"), Jonson's comical Mammon is content to enchant the "sublim'd pure wife" of a local burgess. [*Alchemist*, II: ii] But when it is taken up by William Shakespeare, Jonson's more famous contemporary, the magician-technologist regains his commanding stature and achieves his most sublime characterization.

In *The Tempest*, Shakespeare's Prospero has renounced his kingdom to pursue the sciences that would unlock the secrets of nature and place them at his command:

> those being all my study,
> The government I cast upon my brother,
> And to my state grew stranger, being transported
> And rapt in secret studies.
>
> [*Tempest*, I: 2, 74–7]

Neglecting matters of state—"My library was dukedom large enough," he recalls—Prospero is undone by the machinations of his evil brother, who usurps his kingdom and casts him adrift upon the open sea. A loyal courtier, knowing of the authority that ancient texts held for the magician, "furnished me from mine own library with volumes that I prize above my dukedom."

Accompanied by his infant daughter and armed with his precious cargo of occult books and magical apparatus, Prospero makes his way to the safety of an island whose human and natural resources he subdues by his magic art: the native Caliban, whose labor he effortlessly commands, and the sprite Ariel. This airy servant, who assumes the form of wind and rain and fire, embodies the dream of a fully domesticated nature that will preoccupy the European technological imagination from this point in history onward:

> All hail, great master, grave sir, hail! I come
> To answer thy best pleasure, be't to fly,
> To swim, to dive into the fire, to ride
> On the curled clouds; to thy strong bidding task
> Ariel and all his quality.
>
> [*Tempest*, I: 2, 189–93]

The forces of nature have been rendered wholly tractable and stand in waiting, subject to the whims of their human masters, but longing for their freedom.

Shakespeare had already, through the character of Hotspur in *Henry IV*, warned that "Diseaséd nature oftentimes breaks forth in strange eruptions" and that "the old beldame earth" may exact a terrible vengeance against those who would use her carelessly. [*Henry IV, Part 1*, III: 1, 25–35] This is just the warning that recurs, in allegorical form, in *The Tempest*, which was written late in Shakespeare's life and represents his mature thoughts on technology and the domination of nature. Though they were perfectly capable of depleting their soil, ravaging their forests, and polluting their streams, human populations had never before possessed the technical means with which to lay waste to oceans and ravage whole continents. But in the ethos of the Renaissance magician, who dreamt that the heavens and the earth could be placed at his disposal, the conquest of an adversarial nature began in earnest.

The Tempest has been called "Shakespeare's American fable," for he is known to have been inspired by reports from the New World. For a century, already, availing themselves of new technologies of navigation, cartography, and ballistics, Europeans had been embarked upon their program of global conquest. The Americas, in particular, represented a geographical expanse of untamed nature, but in the reports of the noble savages who dwelt there in a presumed state of innocence, there were hints also of a vast and uncharted psychological landscape. The role of science and technology in taming the forces of both physical and *human* nature had become an acute problem in the European consciousness by the time that Prospero confronted the moral limits of his art.

Prospero, then, is not Faust. At the last possible moment, before his soul is lost for eternity, he steps back from his grandiose scheme. Through his black arts he has raised a tempest and caused his enemies to be shipwrecked on his pristine island, free to remake themselves in the image of uncorrupted nature. But the same Prospero who has "bedimmed the noontid sun, called forth the mutinous winds," set the elements at war with one another and summoned the spirits of the dead, discovers through the petty intrigues of his guests that human nature defies intervention. He resolves to break his magic staff, put off his sorcerer's robes, and cast his books into the sea. Recognizing the limits beyond which his technical powers must not extend, he renounces his "rough magic" and prepares to rejoin the human community. [*Tempest*, V: 1, 33ff.]

For all the time and resources and intelligence expended upon them, the occult sciences failed to produce the results that Prospero and his fellow practitioners had sought: No base metal was ever transmuted into gold; Ben Jonson's Alchemist found no elixir that "by it's vertue, Can confer honor, respect, long life"; and late in his life (he died at the ripe old age of 31), even the esoteric Pico della Mirandola came to denounce the pretensions of the astrologers as "the most infectious of all frauds." The problem was not that the demand for the unification of theory and practice was misguided, but that it was premature. To elevate scientific theory to a point from which it could improve upon empirical, rule-of-thumb techniques would be the task of a different breed of thinker.

In the year 1500, the worldview of an educated citizen began with the presumption that the earth was a motionless globe, resting at the center of ten concentric, crystalline shells in which were embedded the sun and moon, the planets, and the fixed stars. The four terrestrial elements identified by Aristotle (still the ultimate authority on all points of science), obeyed a simple rectilinear physics (earth and water flow downward, air and fire ascend upward), while the heavenly bodies followed their divinely circular paths. Medieval theologians had translated the Unmoved Mover of the pagan natural philosophers into the God of the Christian Church, so that the whole operation was now overseen by a benevolent, intelligent Creator who dwelt beyond the tenth sphere in an ethereal, incorruptible "fifth essence." Mankind was positioned exactly at the center of creation (with womankind slightly off to one side), and the entire cosmos was infused with meaning and purpose.

Two centuries later, this medieval world view lay in ruins. The earth had been thrust by the astronomers from its place of honor and human beings found themselves perched precariously on a wet rock flying through infinite empty space at 100,000 feet per second. God, once grandly depicted as the divine engineer-architect of the universe, was reduced to a humble watchmaker who wound his cosmic mechanism but then stepped backward to let it run according to its own mathematical laws. Anatomists failed to find the seat of the soul, but discovered instead a circulatory system with disturbing affinities to hydraulic pumps. And the new physics projected a world of cold, lifeless quantities, inconceivably vast, unresponsive to human prayers and indifferent to their complaints. For the first time in history, science depicted a universe that was ultimately and essentially meaningless.

The 17th-century theorists who sparked this intellectual revolution are well known, and their achievements have been justly celebrated: the observations and calculations of Copernicus, Tycho Brahe, and Johannes Kepler, which placed the study of astronomy on its modern trajectory; the "experimental philosophers" Galileo, William Gilbert, and the great Isaac Newton (whose predilections for the occult did not prevent him from laying the foundations of modern physics); William Harvey in medical physiology, Robert Boyle in chemistry, and so on. Behind them stood an army of laboratory assistants and instrument makers, patrons and publishers, students and apprentices, institutions and associations, and wives and husbands whose place in the history of science has only recently come to be recognized.

It was inevitable that the question would be raised, sooner rather than later, of the use to which the new science could be put. The labors of the discredited alchemists and astrologers had at least been directed toward useful ends, but what could be the possible utility of Kepler's calculation of the elliptical orbits of the planets or Boyle's observation, in 1662, that the pressure and volume of a gas are inversely proportional to one another?

Of course, the leading scientists of the age had relied heavily upon technology in their work. Galileo eagerly sought out practical advice from Spanish navigators and Venetian artillerymen; Pascal and Leibniz both believed their mathematical problems could be solved by machines; and the pendulum clock, the telescope, and the thermometer testify to the mounting collaboration of the theorist and the instrument-maker. But how was the debt of science to technology to be repaid? The first to answer this challenge was Francis Bacon (1561–1626), the thinker who most clearly grasped that in the new age that was dawning, "those twin objects, human Knowledge and human Power, do really meet in one."

Bacon is the pivotal figure in the changing fortunes of technology and of those who practice it. Throughout the literature of the Renaissance, as we have seen, the idea had been germinating that theory, philosophy, and science might profitably be applied to human life and need not be conceived as ends in themselves. It was Bacon who brought this notion to maturity, rejecting the classical presumption "that the contemplation of truth is a thing worthier and loftier than all utility." This did not mean that the empirical, practical work of the mechanic was of a higher order than that of the theorist, however, but that they must henceforth become one. "We cannot command nature," he foresaw, "except by obeying her." [Organon, I: CXXIV; I: CXXIX]

Bacon's call for the methodical application of science to technology represented a startling break with the past and a profound insight into the future. In a single, sweeping gesture he repudiated the burdensome legacy of antiquity and declared that we are the true "ancients," for the Greek and Latin philosophers lived in the infancy of scientific thought. Claiming that "the end which this science of mine proposes is the invention not of arguments but of arts," he was the first to propose that progress may be measured with a technological yardstick, and that the measure of technology, in turn, was "the relief of man's estate." [*Instauratio*, p. 19]

In arguing for his new science-based technology, Bacon observed that three "mechanical discoveries" had transformed the conditions of human life to an incalculably greater extent than any empire, sect, or star: the magnetic needle, which, enclosed in a compass, gave navigators the confidence to set out across the open sea; the printing press, which revolutionized the propagation of knowledge; and gunpowder, which transformed the conditions of politics and warfare. [*Organon*, I: CXXIX] The origins of these devices, however, were "obscure and inglorious," which led Bacon to wonder what might be achieved if the "inquisition of nature" were organized methodically and systematically. Late in his life he began to imagine the institutional setting in which the organized pursuit of knowledge could be permanently sustained.

In his unfinished *New Atlantis* (1627), Bacon imagined the island civilization of Bensalem, at whose center lay the House of Salomon, an institution dedicated to systematic investigation of nature—not for its own sake, but toward "the enlarging of the bounds of Human Empire." [*Atlantis*, p. 480] His visionary tract was only the latest in a tradition of imaginative fiction that stretched back a century to the *Utopia* of Thomas More. Whereas More had devoted his pages to the political institutions that sustain the contemplative ideal of his Utopians, however, Bacon is fatefully silent about the governance of Bensalem. The marriage of science and technology secretly consummated in the House of Salomon yields only solutions, not problems.

Only today, with our heightened sense of the fragility of the earth and of its human and natural ecologies, have the full implications of Bacon's conception of nature become apparent: its engendering as female, and (as such!!) an object of domination, mastery, and control; and its treatment as an inexhaustible source of raw materials awaiting industrial exploitation. But while there is much in his philosophy that we should not too readily admire, the colossal vision of the man is indisputable. In 1660, not many years after his death, the Royal Society was founded in London "for the Improving of Natural Knowledge." The official patron of this distinguished body was Charles II, but its chief inspiration was acknowledged to be the Baron of Verulam, Lord Francis Bacon.

Other "mechanical philosophers" developed the Baconian world view, even when they dissented from it in important respects. The mathematician René Descartes, whose famous "mind-body dualism" drove a permanent wedge between human consciousness and the mechanical universe, could still imagine "a practical philosophy by means of which . . . we may render ourselves masters and possessors of nature." [*Method,* VI] Marin Mersenne wrote in 1634 that through their statics, hydraulics, and pneumatics, "men can imitate the most admirable works of God." In Germany the philosopher-mathematician Leibniz applied his energies (unsuccessfully) to harnessing the power of of water and wind. And Christiaan Huygens proposed to the French Minister Colbert the design of internal combustion engines actuated by the force of gunpowder, which "has hitherto served only for violent action."

But there were skeptics as well. The mechanistic worldview of the philosophers was not immediately transferable to useful projects, and the manufacturing techniques of early-modern artisans could not yet meet the standards of the new mathematical sciences. Indeed, if the promises of the new experimental science had not so vastly exceeded what, at the beginning of the 18th century, it was capable of delivering, it would be impossible to explain the contemporary popularity of Captain Lemuel Gulliver's voyage to the Flying Island of Laputa.

Borrowing a device from Thomas More, Shakespeare, and Francis Bacon, Jonathan Swift shipwrecks his hapless mariner on an exotic island that reflects in parody the very place from which he had set out. The typical Laputan whom he encounters is an absent-minded fellow, "always so wrapped up in cogitation, that he is in manifest danger of falling down every precipice, and bouncing his head against every post." Gulliver receives a gracious welcome from these exacting folk, however, beginning with a fine dinner consisting of "a shoulder of mutton, cut into an equilateral triangle, a piece of beef into rhomboides, and a pudding into a cycloid." And a few days later, "observing how ill I was clad," the King of the Laputans arranges for a tailor to visit him:

> This operator did his office after a different manner from those of his trade in Europe. He first took my altitude by a quadrant, and then, with rule and compasses, described the dimensions and outlines of my whole body, all which he entered upon paper, and in six days brought my clothes very ill made, and quite out of shape, by happening to mistake a figure in the calculation.

> [*Gulliver's Travels*, p. 130]

Their houses are poorly made as their clothes on account of "the contempt they bear for practical geometry, which they despise as vulgar and mechanic," and their language prevents them from expressing their thoughts except in terms of music and mathematics.

After three years on the island, Captain Gulliver is finally invited to visit the Grand Academy of Lagado, a wicked parody of Francis Bacon's dream of a House of Salomon and its realization in the Royal Society. We know from his journals that Swift had himself toured the Society's quarters at Gresham College (on the same day that he visited the lunatic asylum at Bedlam, and on his way to an evening puppet show). Gulliver's experience reflects both the madness and the theatricality of that day, as well as the outright pretensions of modern science. Corresponding to the "Experiments of Light" and "Experiments of Fruit" that Bacon had envisaged, half of the Academy has been appropriated to "the advancers of speculative learning," the other to the "Projectors" who hope to apply this learning to practical tasks: extracting sunshine from cucumbers, separating human excrement into its edible components, building houses from the roof downward, and softening marble blocks so that they may be used for pillows and pincushions.

Although these extravagant lunacies have been attributed to Swift's undisciplined literary imagination, almost every detail of Gulliver's voyage to Laputa was culled from the *Philosophical Transactions* of the Royal Society, whose President, Sir Isaac Newton, died (fortuitously!) in the year that *Gulliver's Travels* was published. Indeed, if we allow him a healthy measure of satirical license, the outlandish schemes Swift concocted were not so different from what he might have found among his scientific contemporaries. By the early 18th century, educated Europeans were following with passionate interest the applications of inductive and deductive methods to every phenomenon under—and including—the sun. But this was an age less tolerant of the apparent absurdities of science than our own, which recalls that penicillin was discovered by studying bread molds and the gene-splicing technique by observing how bacteria have sex.

In the 17th and 18th centuries, however, the road from scientific illumination to technological progress was not yet paved, and only a few visionaries, like Francis Bacon, had ventured down it. Toward the end of his life Bacon had paused in his labors to outline a monumental, collaborative research program that would be nothing less than the history of nature, insofar as "she is constrained and molded by art and human ministry." The proposed 130 separate "histories" would cover everything from geologic and astronomical phenomena to the tools, techniques, and traditions of artisans ("mechanical and illiberal as it may seem"), so that for the first time technics would be recognized "as streams flowing from all sides into the sea of philosophy." Were this vast, encyclopedic project to be completed, he mused, "then shall we be no longer kept dancing within little rings, like persons bewitched, but our range and circuit will be as wide as the compass of the world." [*Natural History*, I, V, IV]

The encyclopedic impulse gathered momentum throughout Europe in the decades after Bacon's death—it was taken up by

17th-century *virtuosi* such as the English chemist Robert Boyle, the Calvinist philosopher Pierre Bayle in France, and the German mathematician Leibniz, who shrewdly perceived that technology was "the unregistered experience" of the human race. It matured a century later, in the most ambitious intellectual undertaking of the 18th century, the *Encyclopedia* or *"dictionnaire raisoneé" of the sciences, the arts, and the technical trades,* whose 17 volumes of text and 11 remarkable volumes of copperplate engravings were published clandestinely in Paris between 1751 and 1772.

The *Encyclopedia* was overseen by Denis Diderot (1713–84), and its contributors included the leading *philosophes* of the French Enlightenment: the Baron de Montesquieu; the philosophers Voltaire, Helvetius, Condorcet, and the abbé de Condillac; political economists Turgot and Quesnay; the mathematician d'Alembert; and the iconoclastic Jean-Jacques Rousseau. Attempts by the authorities to suppress it—because of its distinctly polemical tone and its dangerously Baconian ethos that "knowledge is power"—ensured its commercial success. It is true that the authors filtered into their separate articles an explicitly liberal ideology; from our point of view, however, the most subversive feature of the French *Encylopedia* was the dignity it accorded to the works of the laboring classes. On the eve of the French Revolution this intrepid band of philosophers ventured out into the fields, factories, and workshops that sustained the *ancien régime* and returned with the insight that the man of letters "does not know the twentieth part" of what moves the modern world.

The writers who conceived and composed the *Encylopedia* aimed at more than a compendium of knowledge: Their program was to heal, once and for all, the breach between scientific speculation and technological practice or, in their own terminology, between the liberal and the mechanical arts. "It is difficult, if not impossible, to go far in the practice of an art without speculation," observed Diderot (himself the son of a cutler), "and, conversely, to have a thorough knowledge of the speculative aspects of an art without being versed in its practice." [*Encyclopedia*, "Art"]

The final reconciliation of head and hand was inhibited on two counts. First, and most easily redressed, was the aura of mystery in which the industrial trades were enveloped—a legacy of the protective guilds of the Middle Ages that was perpetuated by the inarticulate and suspicious tradesmen themselves. Accordingly, "we approached the most capable of them in Paris and in the realm," recounted d'Alembert in his *Preliminary Discourse* to the *Encylopedia*:

> We took the trouble of going into their shops, of questioning them, of writing at their dictation, of developing their thoughts and of drawing therefrom the terms peculiar to their professions . . .

But, continued the philosopher, there are some mechanical arts whose operations are so complex that one cannot speak of them except from experience:

> . . . Thus, several times we had to get possession of the machines, to construct them, and to put a hand to the work. It was necessary to become apprentices, so to speak, and to manufacture some poor objects ourselves in order to learn how to teach others the way good specimens are made.
>
> [Jean le Rond d'Alembert,
> *Preliminary Discourse*, Part III]

For twenty years Diderot and his comrades tramped up and down the French countryside investigating the manufacture of everything from plate glass to playing cards, the operation of canals and cannon foundries, the workshops of wigmakers and wheelwrights. Their first task was one of enlightenment—harmless enough in itself, perhaps, for it could be met by means of *information*. The second obstacle that had to be overcome was a deeply ingrained disdain for the industrial arts, which could be surmounted, Diderot believed, only by "changing the general way of thinking."

"Let us finally render artisans the justice that is their due," he cried in the opening article of the *Encyclopedia*:

Figure 4.2

Place on one side of the scales the actual advantages of the most sublime sciences, and on the other side the advantages of the mechanical arts. . . . You will discover that far more praise has been heaped upon those men who spend their time making us believe we are happy, than on those who actually bring us happiness. How strangely we judge!

Denis Diderot, "Art"

Encyclopedia, "The Pin Factory," from first volume of plates

The liberal arts have sung their own praise long enough; they should now raise their voice in praise of the mechanical arts. The liberal arts must free the mechanical arts from the degradation in which these have so long been held by prejudice.

[*Encyclopedia*, "Art"]

Such a program had political implications as well, for the elevation of the "illiberal," technical trades implied the elevation of those who practiced them. This dangerously egalitarian message was heard by the authorities no less than by the artisans themselves, and Diderot was kept under police surveillance for the duration.

By this time, however, the belief in technological progress was widespread, even if social progress remained an ideal to be fought for. On the eve of the French and the Industrial revolutions the mechanical worldview derived by Descartes and Newton was finding its counterpart in an increasingly mechanized world. The universe was described as a vast clockwork mechanism, while on earth master watchmakers were crafting accurate timepieces barely 15 millimeters in thickness. In 1748 Julien Offray de la Mettrie described not just the human body but the soul, too, as a machine, while artisans such as Vaucanson and Jacquet-Droz toured the courts of Europe with their letter-writing, music-making, and chess-playing automata. The battle for the acceptance of technology, and for linking it at last with the mighty current of mathematical science, appeared to have been won. Only one dissenting voice was raised, but it was a powerful one indeed.

In the summer of 1749, Jean-Jacques Rousseau set out by foot from Paris to visit his comrade Diderot, who had been imprisoned at Vincennes. "The summer of 1749 was excessively hot," he later recalled, and to moderate his pace he began to read from the daily *Mercure de France*:

While reading as I walked, I came upon the subject proposed by the Academy of Dijon as a prize essay for the following year: "Has the

progress of the arts and sciences contributed more to the corruption
or purification of morals?" From the moment I read these words, I
beheld another world and became another man.

[*Confessions*, Bk. 7]

The "other world" beheld by Rousseau was a grim alternative to
the vision of unbounded progress promised by the
Enlightenment, one in which the extension of our powers over
nature were not accompanied by a corresponding increase in
our capacity to exercise this power wisely. In fact, while setting
out to visit its outstanding champion, Rousseau conceived the
most blistering attack against the dream of reason that has ever
been penned. It was in this sense that he "became another man"
than the one projected by the rationalist ideologies of the
Enlightenment.

In his prizewinning essay, known today as the *First
Discourse*, Rousseau recast the terms of the debate over the arts
and sciences: knowledge is not about *power*, he objected, it is about
virtue; and from this perspective it seemed clear that "our souls
have been corrupted in proportion to the advancement of our
sciences and arts toward perfection." The harsh light of reason
had exposed the dogmas of the Church and the ideologies of the
absolutist state, but would the arts and sciences be capable of
forging new bonds of human solidarity in their place? "We have
physicists, geometers, chemists, astronomers, poets, musicians,
painters," he lamented; "we no longer have citizens." [*Discourse*,
p. 39, 59]

The independent spirits of science and technology—the
works of the head and of the hand—had lived apart for most of
human history, and began to flirt audaciously with one another
only in the amorous culture of the Renaissance. There followed,
as is so often the case in adolescent romance, the rapid maturation
of the less-developed partner, and after nearly a century of
scientific revolution a few farsighted thinkers began to see
grounds for a reconciliation. During the Enlightenment the two
began to court one another openly—in the pages of the *Encyclopedia*

and in the meeting rooms of the Royal Society, the *Académie Française*, and other learned settings. On the eve of the Industrial Revolution the imminent marriage of science and technology was being widely announced. Only Rousseau seems to have wondered about the offspring of that union, and to have asked himself whether we had not created a monster.

5

∎ ∎ ∎ ∎ ∎ ▪

Technology and Revolution

"Gentlemen!" exulted G.W.F. Hegel in his
concluding lecture of 1806. "We find ourselves in
an important epoch, in a fermentation, in which
Spirit has made a leap forward. . . . The whole
mass of ideas and concepts that have

been current until now, the very bonds of the world, are dissolved and collapsing into themselves like a vision in a dream." The philosopher had good reason for excitement, for the revolution sweeping across old Europe was destined to transform utterly the face of the continent and the world.

The French Revolution, in whose afterglow the philosopher spoke, had engaged the ideas and emotions of a generation of thinkers: Inspired by its message, Thomas Paine and Mary Wollstonecraft championed the rights of men and women respectively, while the fiery Edmund Burke lamented that "the glory of Europe is extinguished forever"; the news caused the austere philosopher Immanuel Kant to miss his afternoon walk, and Hegel himself proclaimed that "a new emergence of Spirit is at hand;" painters such as Jacques-Louis David in France and Fransisco de Goya in Spain captured its emblems. The French Revolution hides in the libretto of Mozart's *Don Giovanni* and commands center stage in Beethoven's *Emperor* Concerto.

While it is true that the political revolution in France bestowed a new set of symbols upon modern life, it was the Industrial Revolution, whose flames were being fanned across the Channel in England, that transformed the conditions of life itself. The incremental process of industrialization, however, involved no storming of the Bastille or beheading of kings, and it registered first on the European consciousness more in the form of intuitions and impressions than of analyses. More readily in the minds of poets and philosophers than of politicians did the forms of the new order take shape. "Not the external and physical alone is now managed by machines," wrote the young Thomas Carlyle, "but the internal and spiritual also. . . . Men are grown mechanical in head and in heart, as well as in hand." [*Signs of the Times*]

The fortunes of the Faust legend are always a close indicator of the moral dilemmas of an age, and in the hands of Johann Wolfgang von Goethe (1749–1832) it was made to speak directly to the challenges of the industrial era. The composition of *Faust*, which spanned the first six decades of European industrialization, captures the essence of the new human type whose outlines were coming dimly into view. In contrast to his Renaissance forebears, the 19th-century Faust is defined not by some goal toward which he strives but by striving itself. Goethe's Faust is not driven by a desire for wealth or power or even knowledge, but like Mozart's amorous Don Giovanni, by a desire for desire—the one longing that can never be fulfilled.

At the heart of the Faust legend, from Christopher Marlowe to the Rolling Stones, is the *Pakt*, the bargain struck between the doctor and the devil. But in an unprecedented twist to the story, Goethe shifts the terms of the deal. The pact subtly becomes a be*t*, in which Mephistopheles wagers that the learned doctor can be reconciled to the status quo. Faust scoffs at the claim, and bets nothing less than his immortal soul that, alas, no power on heaven or earth can still the restless surging of his heart:

> Should I, to any moment, say,
> 'Linger on, you are so fair!'
> Then may you fasten me in chains,
> Then my ruin I will gladly bear.
>
> [*Faust*, I: 1699–1702]

Goethe's Faust is the very type of the modern technocrat, so utterly confident that whatever is could always be bigger, faster, more powerful, more ambitious, that he will court eternal damnation rather than renounce the vision of infinite progress that possesses him. With Mephisto at his side, cast now as a shop foreman, now as a field lieutenant, Faust schemes to roll back the sea, bring whole continents under cultivation, and conjure cities out of thin air. He is cut from the same cloth as the restless entrepreneurs whom Carlyle christened "Captains of Industry," men whose psychological and technological powers were driving society toward "a nobler Hell and a nobler Heaven." [*Past and Present*, IV: iv]

In Carlyle's England, where the inexorable march of industry was more advanced than anywhere on the Continent, the same fears were being given poetic form. In *Defense of Poetry* itself Percy Shelley warned that the technological impulse must be tempered by spiritual values: "The cultivation of those sciences which have enlarged the limits of the empire of man over the external world, has, for want of the poetical faculty, proportionately circumscribed those of the internal world." [*Defense of Poetry,* p. 69] It was Shelley's young wife, however, who three years earlier had captured most profoundly the image of Prometheus Unbound. In 1818 the 21-year-old Mary Shelley asked whether the dream of reason had not in fact produced a nightmarish monster bent on our destruction; and in the maniacal figure of Victor Frankenstein she created an emblem of the industrial age.

Despite its stark originality, Shelley's *Frankenstein, or the Modern Prometheus* engaged not just the modern predicament but the accumulated cultural legacy of the West. The ancient Prometheus, it will be remembered, was condemned to eternal damnation for having stolen a spark of divine fire; his 19th-century counterpart, cursed with the same fascination, suffered the same fate. The modern Prometheus wished to wrest from the gods not the mysteries of divine fire, but the physical laws of electricity and galvanism. Victor Frankenstein's ambition was to "infuse a spark of being" into the lifeless form he had assembled in his attic laboratory, and in doing so to stand with the ancient Titan among the benefactors of humankind. This modest ambition soon yielded to greater things.

The young Victor Frankenstein, we are told, was driven from an early age to understand the physical secrets of the world. This quest set him apart from his childhood friend Henry Clerval, who "occupied himself, so to speak, with the moral relations of things," and from his beloved Elizabeth Lavenza, who "busied herself with following the aerial creations of the poets." [*Frankenstein,* 36–8] The very word "science"—like its etymological cousins "scissors" and "schism" and "schizophrenia"—implies a dividing up or a splitting off; its practice required that Victor separate himself from all residues of the poetic and moralistic thinking that had impeded the birth of modern science.

In his education, then, Frankenstein must retrace the steps by which science emancipated itself from its magical past. Upon arriving at the university in Ingolstadt he is instructed to renounce the lords of his childish imagination—Cornelius Agrippa, Albertus Magnus, and Paracelsus, whom we met in our encounter with the holistic magical tradition: "The ancient teachers of this science," declares his chemistry professor, "promised impossibilities and performed nothing." The modern masters, by contrast, promise very little:

> They know that metals cannot be transmuted, and that the elixir of life is a chimera. But these philosophers, whose hands seem only made to dabble in dirt, and their eyes to pore over the microscope or crucible, have indeed performed miracles. They penetrate into the recesses of nature, and show how she works in her hiding places. They ascend into the heavens: They have discovered how the blood circulates, and the nature of the air we breathe. They have acquired new and almost unlimited powers; they can command the thunders of heaven, mimic the earthquake, and even mock the invisible world with its own shadows.
>
> [*Frankenstein*, ch. 3]

The frustrated Victor, having left his beloved Elizabeth behind but still agitated by an adolescent longing to "penetrate into the recesses of nature," is initiated here into the doctrine of the Baconian fraternity: Technology can command nature only if science first obeys her.

This is the message imparted by the pregnant teenager whose mother, the great feminist writer Mary Wollstonecraft, had died giving birth to her, and who had already lost one child herself. Preoccupied with the theme of childbirth, Mary Shelley conceived the story of a scientist whose idea gestates for nine months ("winter, spring, and summer passed away") and who then, in a single violent convulsion, gives birth to a creature that,

as the saying goes, 'only a mother could love.' Victor Frankenstein, however, cannot; he recoils in horror and flies from the
room, and his unnatural offspring, bereft of guidance, nurturance,
and companionship, wreaks a terrible vengeance on those who
abandoned him.

The theme of technics-out-of-control looms so large in the
discourse of modernity that it is difficult to appreciate the
prescience of Shelley's insight. She was not the first European
author to warn against the possible abuses of technology—the
Italian humanist Petrarch, having observed the firing of a cannon
in 1350, marveled at how "quick and ingenious are the minds of
men in learning the most pernicious arts," and Leonardo da
Vinci deliberately suppressed his design for a submarine "on
account of the evil nature of men who would practice
assassinations at the bottom of the seas." But Shelley was not
merely warning against the application of technology to violent
or nefarious ends; more clearly than any of her contemporaries,
she perceived the dangers inherent in technology itself.

She would soon be joined by a chorus of other voices. The
system that Carlyle was the first to call "industrialism" had by
this time begun to transform the landscape of the 19th century,
and contemporaries viewed it with a mixture of fascination and
foreboding. The new manufacturing techniques were of a wholly
different character from the small-scale workshops lovingly
illustrated in the plates appended to Diderot's *Encyclopedia* but
doomed by their absence of an external power source. The
revolutionary technologies of the industrial age, by contrast,
were powered by a self-regulating mechanism that had taken
centuries to perfect and that now held out the promise of
unlimited productivity.

This mechanism was not the atmospheric steam engine,
successively improved by Thomas Newcomen, Matthew Boulton,
and James Watt. It was not the hand-powered spinning jenny of
James Hargreaves, the mechanized water frame built by the
wigmaker Richard Arkwright in 1769, or any of the other
inventions that revolutionized the 18th-century textile industry.

It was not the railway locomotive, the cotton gin, or the coal-fired blast furnace. The self-regulating mechanism that powered the Industrial Revolution in Britain, described in 1776 not by an engineer but by a professor of Moral Philosophy at Glasgow, was the market:

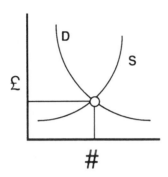

In his *Inquiry into the Nature and Causes of the Wealth of Nations*, Adam Smith set out to show why a modern economy does not need economists to look after it. Left to its own devices, freed from the interference of royal prerogatives, feudal entailments, or manufacturers' monopolies, the market mechanism itself will ensure that suppliers throw their goods onto the market in direct relation to the demand for those goods, such that their price will be determined by the intersection of two curves, one representing demand and the other supply. Fuelled only by the self-interest of individuals (a renewable energy source if their ever was one!), this was a machine that would run by itself more efficiently than all of the perpetual motion machines of the Middle Ages. The egoistic individual, "led by an invisible hand to promote an end which was no part of his own intention," increases the wealth not only of himself but of entire nations. [*Wealth of Nations*, Bk. 4, ch. 2]

Several conditions had to be met before this supposedly "natural" means of economic exchange could begin to function. Most impediments to free trade would have to be dismantled; well-intentioned social policies would have to be abolished, for as Smith continued, "I have never known much good done by those who affected to trade for the public good"; most importantly, the basic factors of modern industrial production—land, labor, and capital—would have to be thrown as commodities onto the market, equally available to anyone who could pay their market price. Adam Smith introduced myriad reservations into his text, but his 19th-century followers, deafened by the grinding of the market mechanism, paid them little heed. Few asked what would happen if this actually came to pass—if the resources of nature, the wealth of nations, and the labor that is (inconveniently) attached to human bodies were actually reduced to the abstract status of commodities to be bought and sold.

Before long the answer began to take shape: Nature would be despoiled to a degree that not even Milton had imagined in *Paradise Lost* when he described, in the building of Pandæmonium, how men, "with impious hands, Rifl'd the bowels of their mother Earth for Treasures better hid;" human beings would be exploited beyond endurance in the "dark Satanic mills" that haunted the imagination of William Blake. A few classical Liberals (today they would be called conservatives) answered that so long as nothing is done to interfere with the machinery of the market, in the long run it will recover its equilibrium. Most analysts recognized, however, that life is lived in the short run, and in the short run the very sources of industrial wealth were being systematically annihilated. The monstrous creation that was wreaking such a terrible vengeance was not a green-skinned "Frankenstein" with watery eyes and a bolt sticking out of his forehead—it was a two-armed graph labeled 'supply' and 'demand'—but it issued the same inflexible command: "You are my creator, but I am your master—obey!"

Driven by the insatiable fury of the market, the first spasms of industrialization produced unimaginable horrors: the pauperization of vast sectors of the agricultural and industrial labor force; the destruction of families, communities, and traditional social bonds; the virtual enslavement of children; the growth of filthy, disease- and crime-ridden slums; and everywhere a widening gulf between the newly rich and the newly destitute. The indifference of the capitalist bourgeoisie before this spectacle of misery was preserved in a memorable account left by Friedrich Engels:

> One day I walked with one of these middle-class gentlemen into Manchester. I spoke to him about the disgraceful unhealthy slums and drew his attention to the disgusting conditions of that part of the town in which the factory workers lived. I declared that I had never seen so badly built a town in all my life. He listened patiently and at the corner of the street at which we parted company he remarked: 'And yet there is a great deal of money made here. Good morning, Sir.'
>
> [*Condition of the Working Class*, ch. 12]

Even before the outrages of the English factory system began to be documented—in the findings of government commissions, the petitions of social reformers, the novels of Dickens and Disraeli and Elizabeth Gaskell—efforts were made to blunt the hard edge of the machine. The utilitarian radical Jeremy Bentham (1748–1832) argued to the end of his life for his imagined "Panopticon," an enlightened penal institution that would replace the blows of the jailer with the ubiquitous eye of the observer. In Scotland the philanthropic industrialist Robert Owen created a model factory town at New Lanark that was visited by enlightened dignitaries from across Europe. The liberal John Stuart Mill questioned whether "all the mechanical inventions yet made have lightened the day's toil of a single human being," [*Principles*, IV: ch. 6] and argued for a moderate program of reform.

Across the Channel, where the march of industry was less unrelenting than in England, visionaries struggled to retain society's mastery over the machine. In Paris the eccentric Comte de Saint-Simon, arguing that "scientists, artists, and leaders of industrial enterprises are the men who should be entrusted with administrative power," proposed a utopian *"systéme industriel"* that promised to safeguard the rights and the dignity of the working class. [*Social Organization*, pp. 78–9] In provincial Lyon, Charles Fourier conjured in his imagination fantastic communities in which the depredations of "civilized industry" would be replaced by a libidinally fulfilling system of "passionate attraction."

While the so-called "Utopian Socialists" were uncertain about the mechanization of industry, they shared a reaction against the mechanization of individual and society. But the process seemed, if not inevitable, then at least irreversible. During the first half of the 19th century, technological innovations multiplied at a pace unprecedented in human history: the railroad and the iron-hulled steamship; telegraphy and photography; the factory system and standardized machine tools. Some 8,487 patents were granted in Britain alone in the thirty years between the publication of *Frankenstein* (1818) and the *Communist Manifesto* of 1848.

Eighteen forty-eight marks a symbolic turning-point, for in the first months of that year, in a frenzy of revolutionary activity, the accumulated rage and frustration of two generations burst against the walls of old Europe. Undoubtedly, the profoundest expression of the revolutionary spirit was the pamphlet circulated by an obscure organization that styled itself the Communist League. With mounting pique the officers of the League's Central Committee awaited the position paper they had commissioned from a radical journalist living in exile in Belgium. Finally they lost patience and directed the District Committee of Brussels "to inform Citizen Marx that if the Manifesto of the Communist Party which he agreed, at the last congress, to draw up, does not reach London before Tuesday, February 1st, further measures will be taken against him." Karl Marx complied, and in collaboration with Friedrich Engels produced a revolutionary classic that is at once a fiery call to arms and a sober analysis of the politics of technology.

Marx and Engels argued that the modern capitalist cannot withstand the competitiveness of modern industry without constant technological innovation. Thus the owner of the factory, mill, or mine no matter how reactionary his political views, stands as a representative of the most revolutionary class that has ever trod upon the earth:

> The bourgeoisie, in its rule of scarce one hundred years, has created more massive and more colossal productive forces than have all preceding generations together. Subjugation of nature's forces to man, machinery, application of chemistry to industry and agriculture, steam navigation, electric telegraphs, clearing of whole continents for cultivation, canalization of rivers, whole populations conjured out of the ground . . .
>
> [Communist Manifesto, ch. 1]

The capitalist bourgeoisie, marveled these two radical communists, with its unheard-of productive capacities, has forged a world market, created a world literature, and rescued a considerable portion of humanity from "the idiocy of rural life"; "it has accomplished wonders far surpassing Egyptian pyramids, Roman aqueducts, and Gothic cathedrals."

More than any of their contemporaries, Marx and Engels appreciated the physical and psychological suffering caused by the new techniques and the factory system they required. But their revolutionary hatred for the agents of industrialization did not extend to the technologies of modern industry. They believed that the "alienation" of the laborer in modern industry was not a function of the machine *per se*, but of the deployment of the machine under conditions of capitalist production.

Technology, when subjected to dialectical scrutiny, proved to be a double-edged sword. When it rests in the private hands of a capitalist it serves as an instrument of exploitation by increasing the productivity of the working class without increasing its wages; democratized, wielded by society as a whole, it could be the instrument of liberation. Far from being the culprit, Marx and Engels insisted that the full development of society's technological capabilities was the essential condition of liberation: "Slavery cannot be abolished without the steam-engine and the spinning-jenny, serfdom cannot be abolished without improved agriculture. . . . people cannot be liberated as long as they are unable to obtain food and drink, housing and clothing, in adequate quality and quantity." [*German Ideology,* p. 61]

It would be another twenty years before Marx completed the detailed analysis of machinery and large-scale production that appears in *Das Kapital.* In the interim, events had taken a dramatically different turn from what he and Engels had confidently predicted in 1848. The wave of insurrection that co-incided with the publication of the *Manifesto* briefly threatened Europe's monarchies and bourgeois republics, but was turned back. Recognizing that the industrial proletariat had not matured as rapidly as the industrial technologies that had called it forth, Marx retired to a life of study in the British Museum to analyze the reasons for defeat and the prospects of success.

Meanwhile, the triumphant bourgeoisie staged a massive party to mark the victory of its own revolution. In 1851 Queen Victoria issued an invitation to the nations of the world to come to London to display, "in a spirit of friendly rivalry," the finest products of their factories, mills, and workshops. All that was required was a building suitable to an International Exhibition of the Works of Art and Industry of All Nations, but for this there was no architectural precedent. Some 245 designs were submitted from all parts of Britain, and each was reviewed and duly rejected by a blue-ribbon panel that included the railroad engineer Robert Stephenson; Isambard Kingdom Brunel, builder of the first transatlantic steamships; and architect Charles Barry, designer of Britain's new Houses of Parliament. In a display of lofty self-sacrifice these worthies finally decided to design their own edifice, and soon proposed a monstrous hybrid that resembled (not surprisingly) a cross between a railway tunnel, an immense boiler, and Westminster Hall. They were rescued by a gardener.

Joseph Paxton had never built anything larger than a green-house in which to shelter the prize orchids of his employer, the Duke of Devonshire. Intrigued by the challenge, however, he

Figure 5.1

It may be said without presumption, that an event like this Exhibition could not have taken place at any earlier period, and perhaps not among any other people than ourselves. The friendly confidence reposed by other nations in our institutions; the perfect security for property; the commercial freedom, and the facility of transport, which England pre-eminently possesses, may all be brought forward as causes which have operated in establishing the Exhibition in London.

Official Catalogue of the Great Exhibition
of the Works of Industry of All Nations, 1851

Paxton's initial sketch for the Crystal Palace

sketched a design for a vast structure to be built, like his greenhouse at Chatsworth, entirely of iron and glass. The panes, beams, and girders would be standardized, pre-fabricated, mass-produced—and (as we now say) recyclable, since the exhibition hall was to be dismantled immediately following the Exhibition so that Hyde Park could be restored in its pristine beauty to the City of London. Six weeks after construction began the "Crystal Palace" stood complete; critics fumed that by its absence of ornament or historical reference it resembled a work of engineering more than a work of architecture, and they were right.

Six million visitors flocked to the International Exhibition to view the attempts of modern industry "to wed high art with mechanical skill." With few exceptions, what they saw represented grounds for a divorce: machine-made products smothered in baroque curlicues; frivolous expressions of middle-class affluence; dubious inventions such as the "silent alarm clock"; and heavy machinery adorned with Egyptian pilasters, Corinthian columns, and Gothic arches. For its mechanical principles industry looked to the present and the future; for its aesthetics it was mired in a hopeless, anachronistic search through the past.

Into this confused and contradictory picture marched the indomitable figure of William Morris, who decried the degradation of both art and artisan in the age of mechanical mass-production. Indifferent to theory but moved by an intense feeling for humanity, Morris gazed backward across the centuries and asked whether it was all to end "in a counting-house on top of a cinder-heap." Defiantly, and "with the conceited courage of a young man," he opened in 1861 the firm of "Morris & Co., Fine Art Workmen in Painting, Carving, Furniture, and the Metals," and opened at the same time a new chapter in the the reception of the machine.

In a brief autobiographical sketch Morris described his mission with arresting clarity: "Apart from the desire to produce beautiful things, the leading passion in my life has been and is hatred of modern civilization." ["How I Became a Socialist"] His unshakable conviction was that these two impulses could be harmonized by recovering the guild tradition of the Middle Ages. In a spirit of medievalizing romanticism Morris set about to design a line of hand-crafted products—furniture, wallpaper, typefaces, fabrics, utensils, glassware, and even architecture—that would exemplify the highest standards of craftsmanship. The artist and the artisan would be reunited, and people would be surrounded by things of beauty rather than the shoddy excrescences of the machine.

But the program that was announced by William Morris and the "Arts and Crafts Movement" was a political as well as an aesthetic one, for the prejudice against the "lesser arts" was also a class prejudice against the artisan. This, too, would be overcome. The line between the fine and the applied arts would be blurred, and the "useless toil" imposed by the factory system would become "useful work."

Alas, in the modern age Morris's two driving passions—his love of beauty and his hatred of the machine—proved incompatible. Throughout his life he insisted that "I do not want art for a few, any more than education for a few, or freedom for a few." Yet his workshops turned out fine products that were prized not by the workers who attended his evening lectures on socialism, but by affluent collectors who could afford them. For better or for worse, the machine was here to stay; the task at hand was not to renounce it but to come to terms with it.

Despite the putative failure of the Arts and Crafts movement, William Morris imparted to posterity three interconnected insights that have proven critical to design in the machine age: the conviction that no object is too humble to become a thing of beauty; the idea that good design must integrate form, material, and ornament; and the insistence that a product must meet not only functional but also social and aesthetic standards. Morris envisioned a technology infused with both an ethic and an aesthetic. But it would fall to the next generation, which had reconciled itself to the machine, to realize the program so boldly put forth by William Morris: "redesigning the world."

The 19th century opened in a mood of anticipation and a sense that a profound transformation of human affairs was underway. Hegel likened the new age that was dawning to "a birth-time, and a period of transition [in which] the spirit of man has broken with the old order of things hitherto prevailing." Thomas Carlyle warily surveyed the emerging industrial landscape for "signs of the times," and John Stuart Mill tried to lay hold of "the spirit of the age," though he recognized it to be a moving target. As the new industrial order unfolded, scarcely any artist, intellectual, or official failed to confront it.

The force and ubiquity of technological change, and its impact upon every aspect of life, came to permeate the culture of the later 19th century. It is suggested by the soft images of Manet, as if his misty impressionist landscapes were viewed through the smoke and steam and haze of the industrial city. In America the din of the locomotive—"Type of the modern—emblem of motion and power—pulse of the continent"—can be heard in the poetry of Walt Whitman, and the fragmented rhythms of the assembly line are reflected in the serial photographs of Eadweard Muybridge. In the last quarter of the century a cluster of inventions—the incandescent light bulb and the telephone, the automobile and the streetcar, the wristwatch, the phonograph, the machine gun—disrupted and redefined the prevailing sense of time and of space.

Increasingly, technology came to be seen not just as a fund of new gadgets but as a force that had to be reckoned with theoretically. Marx surely perceived this when he addressed, in an oft-quoted aphorism, the centrality of technology in history: "the handmill gives you society with the feudal lord, the steam-mill society with the industrial capitalist." For Auguste Comte, founder of the positivist philosophy he named "sociology," the true impact of technology lay not so much in past history but in its future reconciliation with science. Comte was optimistic, for according to his law of mental development, mind had lately passed from the theological to the metaphysical into its third and highest stage, the scientific. Although a prodigious labor lay ahead, between the scientist and the industrialist "an intermediate class is rising up—that of the engineers, whose particular function is to settle the relations between theory and practice." [*Positive Philosophy*, p. 41]

The naturalist Charles Darwin went even further, implicitly raising the possibility that through our technologies we may counteract the blind force of natural selection and take control of the evolutionary process itself. The person who puts on a pair of glasses, straps a revolver to his side, or gulps down a tranquilizer has defied in some small measure the tendency of Nature to favor the hardiest specimens among us. This sort of meddling was anathema to his "Social Darwinist" epigones, but they were routed before the end of the century by Darwin's champion, T.H. Huxley: Precisely such interventions, "directed not so much to the survival of the fittest, as to the fitting of as many as possible to survive," are the key to social and ethical progress. [*Evolution and Ethics*, p. 326]

Hegel had cautioned at the beginning of the century that not until the new order had fully matured could it be apprehended philosophically, for "the Owl of Minerva spreads its wings only with the falling of the dusk." By the end of the century, however, the lineaments of the new order were apparent, and the qualified optimism of a Marx or a Comte began to yield to the dark foreboding that would be so characteristic of modern times. The progenitor of the modern worldview was undoubtedly Friedrich Nietzsche, whose death in 1900 signaled what he himself would have called a *Dämmerung*—the twilight of one age and the dawn of another.

Nietzsche's thought, in contrast to the whole tendency of the 19th century, defies categorization. He abhorred intellectual system-building and taught that we must learn "how to philosophize with a hammer." Only rarely did he give voice directly to the specifically technological character of modern culture, and then mostly to denounce "our whole attitude toward nature, the way we violate her with the aid of machines and the heedless inventiveness of our technicians and engineers."

[*Genealogy*, III: para. 9] Yet there is a consistency to him nonetheless, even if it speaks through his brilliant aphorisms more as a mood or a style than as a theory.

At the heart of Nietzsche's thought is the enigmatic concept of the "Will to Power"—an idea that would be savagely corrupted by fascist intellectuals in the 1930s. All life, he believed, was governed by a perpetual drive for "self-overcoming," for going beyond its present state. Scientific knowledge, the pursuit of the constant and the calculable in nature, thus works as a tool of power: "The measure of the desire for knowledge depends on the measure to which the will to power grows in a species: a species grasps a certain amount of reality in order to become master of it, in order to press it into service." [*Will to Power*, III: para. 480]

But what is the logical end to this process, to the perpetual expansion of our knowledge of and power over the universe that seems to be the emblem of the technological imagination? That the will to power had become its own end seemed to Nietzsche to be the characteristic of European civilization in its final, "nihilistic" stage: "What does nihilism mean? *That the highest values devaluate themselves.* The aim is lacking; 'why?' finds no answer." [*Will to Power*, I: para. 2]

Nietzsche, the philosopher of holy madness, in 1889 himself descended into the black night of insanity. His demise was foretold in one of his most powerful passages: "Have you not heard of that madman who lit a lantern in the bright morning hours, ran to the market place, and cried incessantly, 'I seek God! I seek God!'" :

> The madman jumped into their midst and pierced them with his eyes. 'Whither is God?' he cried; 'I will tell you. *We have killed him*— you and I. All of us are his murderers. But how did we do this? How could we drink up the sea? Who gave us the sponge to wipe away the entire horizon? What were we doing when we unchained this earth from its sun? Whither is it moving now? Whither are we moving? Away from all suns? Are we not plunging continually? Backward, sideward, foreward, in all directions? Is there still any up

or down? Are we not straying as through an infinite nothing? Do we not feel the breath of empty space? Has it not become colder? Is not night continually closing in on us? Do we not need to light candles in the morning? Do we hear nothing as yet of the noise of the gravediggers who are burying God? Do we smell nothing as yet of the divine decomposition? Gods, too, decompose. God is dead. God remains dead. And we have killed him.

[*Gay Science*, para. 125]

No single interpretation can illuminate the parable that has haunted the imagination of all succeeding generations; it is enough, perhaps, simply to immerse ourselves in its imagery. Nietzsche's madman seems to recoil in horror at the powers unleashed by the new industrial culture—to drink up the sea, as it were; to wipe away the horizon. But where will we be once we have unchained the earth from its sun, if not plunged into an illimitable darkness? Scientific analysts—"microscopists of the soul," he called them, "vivisectionists of the spirit"—have penetrated every mystery and rendered the universe transparent; armed with technologies of untold power they have dared to usurp the powers of God, have dethroned and even killed him. But are they—are we—big enough to stand alone in a world so wholly of our own making? "There has never been a greater deed," Nietzsche prophesied; "and whoever is born after us—for the sake of this deed he will belong to a higher history than all history hitherto."

This is the desperate challenge that Nietzsche, in his final moments of lucidity, flung at the feet of the 20th century: Will we, who can bring forth new forms of life, who can soar into the heavens and peer into the beginning of time, will we become the new human types who alone can survive our own terrifying powers? Will we, "for the sake of this deed . . . belong to a higher history than all history hitherto"?

6

■ ■ ■ ■ ■ ■

The Prosthetic God

It is a short journey from the incurable
madness of Nietzsche to the incurable sanity of
Sigmund Freud. Nineteen hundred, the year
of Nietzsche's death, was also the year of *The
Interpretation of Dreams*, in which Freud

set out to rival the philosopher, who, he said, "had a more penetrating knowledge of himself than any other man who ever lived." In his early psychoanalytical writings Freud had peered inward at the psychological mechanisms of the self; but gradually, as the terrifying events of the century began to weigh upon his own psyche, he turned outward to a diagnosis of the collective mental disorder called "civilization."

In this bold extrapolation, modern technology played a larger role than has usually been noticed. The highly problematical achievement of civilization, for Freud, resides in the powers of human beings to free themselves from the blind dominon of nature. Every tool, process, or structure enlisted in this campaign thus serves to compensate for the relative deficiencies of the human body: The water wheel, the steam engine, and the electric motor amplify our muscles; our organs of sight and hearing are perfected by telescopes and telephones; the fallibility of memory is corrected by the printed page, the phonograph, and the camera. Had Freud lived only a few years longer he would have approved of the pioneers who called the computer an "electronic brain."

The function of technology, in this view, has been to close the gap between the feeble endowment with which we enter the world and the omniscience and omnipotence we have invested in our gods. In modern times, in which the realm of forbidden longings and unattainable desires has shrunk, this gap has become very narrow indeed. Decked out in all his auxiliary organs—chemical, electrical, mechanical—"man has, as it were, become a kind of prosthetic God." [*Civilization*, ch. 2] Yet the human condition grows ever more precarious.

The explanation, Freud believed, is that we are set upon by three sources of suffering. Like all other machines, our bodies age and decay, causing us inevitable pain and anxiety; second, there is the external world, "which may rage against us with overwhelming and merciless forces of destruction;" finally, our relations with other people are a perpetual source of grief, ranging from the misery of unrequited love to outright acts of violence, prejudice, and general human mendacity. The latter are particularly vexing, for whereas our own mortality and the capriciousness of nature are rooted in the order of things, the depredations of our fellow human beings seem almost gratuitous.

By the 20th century the conclusion was inescapable that the "natural" sources of our discontent—the frailty of our bodies and the capriciousness of nature—were finally yielding to concerted human effort. The practices of modern medicine, the operation of chemical drugs, the measures taken to promote public health, had in Freud's lifetime greatly extended the lifespan allotted to human beings and reduced their susceptibility to disease. On the larger scale, the advance of mathematical science had increased their predictive power over nature, while their technologies had finally begun to control it. "Men are proud of those achievements," Freud conceded," and have a right to be. "But they seem to have observed that this newly won power over space and time, this subjugation of the forces of nature which is the fulfillment of a longing that goes back thousands of years, has not increased the amount of pleasurable satisfaction which they may expect from life and has not made them feel happier." [*Civilization*, ch. 2]

Freud ventured two sorts of explanations for the failure of science and technology to increase our happiness, even as it increased our power over the apparent sources of unhappiness. First, the doctor suggested that technology is, as it were, "iatrogenic," an induced symptom of the ailment it presumes to cure. Despite our illusions of progress, our technologies do nothing but remedy conditions created by prior technologies: Freud can hear the voice of a distant friend over the telephone, but only because the railway had first carried his friend abroad; we can today boast of victorious campaigns against this or that disease, only to return to the cities and workplaces that made us sick in the first place. Technology, Freud concluded, may be likened to "the enjoyment obtained by putting a bare leg from under the bedclothes on a cold winter night and drawing it in again." It is not a majestic story of progress, but of a dog chasing its own tail.

There is a second, even more unsettling reason for the static quality of human existence in the face of such scientific and technological dynamism. Travelling almost unaccompanied through an uncharted psychological wilderness, Freud speculated that even if the subjugation of external nature were achieved, the demands of internal nature—above all the irreducible instinct toward aggression and destruction—would remain powerful enough to deny to us the fruits of our victory. The intractability of what he dared to name the "death instinct" left Freud with the suspicion that here, too, in our unalterable psychic constitution, lies "a piece of unconquerable nature" that no technology can correct.

With a touch of Viennese insouciance, Freud once remarked that the best that psychoanalysis could promise was "transforming your hysterical misery into common unhappiness." [Studies, p. 305] Indeed, despite the morbidity of some of his conclusions, Freud's theories were positively playful in comparison with the dark ruminations of his German contemporaries. The contrast was best captured by the Viennese journalist Karl Kraus, who quipped, in the chaotic aftermath of the First World War, that "in Berlin the situation is serious but not hopeless; in Vienna it is hopeless but not serious." A gloomy forboding hung over the whole spectrum of German thought in the first decades of the 20th century, and nowhere is it more manifest than in matters technological.

This may be explained in large measure by the rapid pace of industrialization in Germany. Whereas in Britain the dislocations of the new industrial order were dispersed over the whole of the 19th century, in Germany the process was compressed into barely a generation, forged in blood and iron, that came of age during the First World War. Massive population shifts from rural areas to the new industrial cities, the disruption of traditional social bonds, the intensification of class conflict and the consequent emergence of mass political parties, and of course the defeat of German arms all contributed to a pervasive "cultural despair" that was fueled by the new industrial technologies.

The thinker who most vividly evokes the spiritual ambience of the era is Oswald Spengler, whose massive inquiry into *The Decline of the West* was published in 1918, at the very moment of Germany's defeat. Spengler elaborated a speculative philosophy of history whose objective, using a method he called "comparative morphology," was to determine nothing less than the destiny of Western Culture. "Culture" he defined as the spiritual core of a people that unites its politics and law, its art and architecture, its religion and philosophy—and its science and technology.

Modestly, Spengler claimed that he had achieved a "Copernican Revolution" in historical study. The original Copernicus, it will be recalled, had dislodged the earth from its privileged place at the center of creation, and Spengler claimed to have done the same for the study of culture. There is no hierarchical succession from "ancient" to "medieval" to "modern," he argued; nor does the culture of the West hold some privileged position with respect to those of other times and other places. In the perspective of world history, cultures are "separate worlds of dynamic being," discrete spiritual entities that "grow with the same superb aimlessness as the flowers of the field." [*Decline*, I: 18, 21] History is not the scene of progress, in this view, but of natural cycles of growth, maturation, and decline. Where, then, lies the character of the present era? What stage in the invariable life-cycle of cultures have we reached?

For Spengler, who viewed himself as the heir not only to Copernicus but to Goethe as well, the soul of Western Culture is fundamentally "Faustian," and its unifying symbol is "pure and limitless space." In contrast to the serene equipoise of classical forms, the orientation of Western Culture for the last one thousand years has been one of thrusting forward into an infinitely receding horizon. Spengler sees the unbounded aspiration of the Faustian soul in the soaring spires of Gothic architecture and in the invention of perspective drawing, which turns the canvas into a window onto infinity; it is suggested by Galilean dynamics and the infinitesimal calculus, which express "the same adamantine will to overcome and break all resistances of the visible"; it is evident in long-distance trade and the long-range weapon. The symbol of infinite, empty space entered the political sphere in the imperial ambitions of European leaders, and is manifest psychologically as "a tense restraint in the void without limits." [*Decline*, I: pp. 183–6]

Even though this Faustian passion altered the face of the earth, Spengler insisted that Western Culture must be viewed as an isolated historical episode, "strictly limited and defined as to form and duration." Moreover, everything about the present era—the widening gap between the productive land and the consuming city, the encroachments of a relentless commercialism, the exhaustion of painting, music, and philosophy—is symptomatic of the passage of the West into its "autumnal" stage: "late, futureless, but quite inevitable." [*Decline*, I: pp. 34–9] Like Nietzsche before him, Spengler saw the nihilism of his age as a symptom of its fatal illness. Only one feature of this grim modern landscape commanded his excitement: the machine. "To me," he confessed, "the depths and refinement of mathematical and physical theories are a joy. . . . I would sooner have the fine, mind-begotten forms of a fast steamer, a steel structure, a precision-lathe, the subtlety and elegance of many chemical and optical processes, than all the pickings and stealings of present-day 'arts and crafts,' architecture and painting included." [*Decline*, I: pp. 43–4]

European technology first subjected nature to its yoke, and then, having become "the real queen of the century," called forth a disciplined human retinue to tend it: the entrepreneur, the engineer, and the factory worker. Driven by his ambition "to break all records and beat all dimensions," the Faustian man thus finds himself enslaved to his own creation: "His number, and the arrangement of life as he lives it, have been driven by the machine on to a path where there is no standing still and no turning back." [*Decline*, II: pp. 499-506] This was no pure celebration of technique, but a desperate appeal to the one force that might survive the demise of a genuine national culture.

What Spengler sought was a Faustian pact with the diabolical machine.

Though Spengler himself remained above the political fray, his ideas resonated ominously with the mood of romantic nationalism that followed hard upon the German defeat. Whether despite or because of its nihilist message, *The Decline of the West* became an instantaneous bestseller and served as an intellectual rallying point for those "reactionary modernists" who saw technology both as the cause of the the present crisis and the means to resolving it.

A sober rejoinder to this sort of technological romanticism was delivered by the sociologist Max Weber, the most powerful analyst of his generation. Weber recognized that the spread of industrialism had brought with it undreamed-of productivity, but that it had also disrupted traditional social bonds and diffused a corrosive "technological rationality" throughout every sphere of life. The calculating rationality, bureaucratic administration, and "mechanized petrification" of life were not incidental side effects that could be willed or wished or legislated away; they were the very essence of industrialism itself. Technology and science had displaced magic and metaphysics; but the resultant "disenchantment of the world" had ushered in a bureaucratized society peopled by "specialists without spirit, sensualists without heart." [*Protestant Ethic*, p. 182] Weber, too, ended his studies on a note of cosmic pessimism: "Not summer's bloom lies before us," he prophesied in his last lecture, "but a polar night of icy darkness and hardness." [*Politics*, p. 128]

Max Weber had traced the roots of capitalist industrialization back to a worldly "Protestant Ethic" that had been ascendant in the West since the 16th century. His contemporary, the conservative economist Werner Sombart (1863–1941), arrived at a rather different conclusion. For Sombart it was not Protestantism but the Jews who transmitted the spirit of capitalism, and with it the dominion of commerce over technology, the entrepreneur over the engineer, and money over the machine. Although he did not go so far as to claim that predatory capitalism was biologically grounded in the Jewish people (non-Jews could equally be infected with it), Sombart's theories lent academic respectability to the quest for a peculiarly German route to modernity that was to culminate in the catastrophe of National Socialism.

The critique of the technological society gathered force and urgency as the clouds of war gathered over Europe, but it was by no means dominated by theorists of the center and the right. From Frankfurt a brilliant group of leftist scholars began in the early 1930s to elaborate a theory of the emergent culture of technology and the forms of thought that corresponded to it. By the time they were driven into exile by the rise of the Nazis, they had laid the foundations for a critical theory of the technological society. Their analysis has had a pervasive influence on both sides of the Atlantic.

The "Frankfurt School," as the group around philosopher Max Horkheimer became known, argued that modern society has inherited from the Enlightenment a conception of rationality that is essentially "instrumental." "This type of reason," wrote Horkheimer," is essentially concerned with means and ends, with the adequacy of procedures for purposes more or less taken for granted." It tends, in other words, to be preoccupied with the most efficient procedures for realizing objectives that are not themselves subject to rational analysis. Like the bureaucrats who engineered Hitler's Final Solution, this type of rationality "attaches little importance to the question whether the purposes as such are reasonable." [*Eclipse of Reason*, p. 3]

In a series of daring explorations, members of the Frankfurt School tracked the reduction of reason to a mere instrument across the whole range of modern culture: Horkheimer and Herbert Marcuse unraveled leading concepts of philosophy to demonstrate how the domination of nature spread to the domination of human beings; Theodor Adorno found evidence in the stupefying rhythms of modern music, and Leo Lowenthal searched out modern literature for signs that art, too, had lost its critical function; from his clinical observations the psychoanalyst Erich Fromm diagnosed the modern fascination with technology as a form of "necrophilia," the transfer of affection from living beings to dead or lifeless things.

A particularly important contribution was made by Walter Benjamin, who analyzed the transformations of perception itself in the age of photography, film, and radio: "The equipment-free aspect of reality has become the height of artifice," he concluded; "the sight of immediate reality has become an orchid in the land of technology." ["Work of Art," p. 233] This was less a wistful longing for a pre-technological age than a demand that the arts unfold in creative interaction with the limits and possibilities of an era of mechanical reproduction. Benjamin died on the Spanish border attempting to escape from Nazi-occupied France, but many of his colleagues made the successful flight to the United States, where they were able to resume an inquiry into technology that had been cut short.

While theorists of left, right, and center were grappling with the politics of technology, another German thinker was attempting to grasp, at the most fundamental level, its philosophical *essence*. Martin Heidegger has become probably the most influential—and surely the most controversial—thinker of the 20th century. His outspoken advocacy of the Nazi Party in the year of his Rectorship of Freiburg University, 1933–34, has utterly discredited him in the eyes of many philosophers, and nothing can ever erase this crime against *philosophy itself*. At the same time, Heidegger's philosophical engagement with technology is of such significance to modern thought that it extends beyond his personal disgrace.

Heidegger undertook a critique, not just of technology, but of an entire tradition of thought that culminated in our present technological world view. The West has evolved a view of nature as a body of raw materials standing "in reserve" and available, even intended, for human use. We are thus the unwitting heirs to a form of humanism derived from the ancient doctrine that "man is the measure of all things." Our needs, our desires, and our capacities "enframe" the world and determine what is to be done with its resources. It follows that modern technology does not impose upon us a given world view but is rather the *product of* an essentially technological world view, one that has been gathering force since the time of the Greeks. "Our age is not a technological age because it is the age of the machine," wrote Heidegger in one of his characteristically enigmatic reversals; "it is an age of the machine because it is the technological age." [*What is Called Thinking?* p. 24]

The framework of Heidegger's questioning of technology was a sustained inquiry into the nature of being—not the be-ing of any particular class of things that may exist, but of Being itself. To treat the external world as a mass of resources put there for our use is to lose sight of their "authentic" mode of Being.

In a rare attempt to render his ideas more concrete, Heidegger considered the essential difference between an old wooden bridge over the Rhine and a modern hydroelectric plant: the former, he concluded, leaves the river to reveal itself in its natural state; the latter, by contrast, sees it only as a potential source of power and does not hesitate to dam it up or redirect its course. Whereas the river defines the bridge, it is the power station that in effect defines the river—in terms of the human uses to which it may be put. [*Question Concerning Technology,* p. 16] The wanton destruction of natural ecologies is one consequence of this misplaced humanism; and our present talk of "human resources" suggests that we have come perilously close to seeing not only nature but ourselves in terms of a "productionist metaphysics."

Heidegger turned instead to the ancient view of technology, buried in the Greek word techne, which embraced both the fine arts and the productive, technical arts. Herein lay the hopeful message that the "makings" of the poetic imagination might yet be reunited with the "makings" of those who practically transform the earth. The search for a mystical union with the life-force of the people—had driven him into the hands of the Nazis. By the 1950s, Heidegger himself understood that this was a dead, and a deadly, idea. He turned instead to the poet Hölderlin, as if to signal that the romantic rejection of technology must yield to a creative appropriation of it:

But where danger is, grows
the saving power also.

Heidegger was by no means alone in his search for the creative source from which both art and technology spring. By the turn of the century, writers of poetry and fiction had begun in earnest to take up the theme of technology, whose symbol was no longer the "village smithy" of Longfellow's New England, but the towering, forty-foot dynamo whose presence was felt by the young Henry Adams "much as the early Christians felt the Cross." [*Education*, ch. 25] In Europe and America the images of technological modernity had registered profoundly upon the literary imagination. The exotic machines of Jules Verne rocketed readers to the far side of the moon and plunged them into the depths of the sea, while the time machine of H.G. Wells sped them far into the future. Back on Earth, Emile Zola descended with them into the coal mines of northern France, and Upton Sinclair led them through the mechanized jungle of the Chicago stockyards.

More venturesome were the "dystopian" novels that began to appear in the 1920s, in which the worst tendencies of the technological society were magnified, generalized, and extrapolated into the future. In dystopian fiction, technology occupied center stage. George Orwell peered ahead to the nightmarish world of 1984, and Aldous Huxley to the "brave new world" of the year 632 A.F. (After Ford), in which, as the World Controller explains, "the machine turns, turns, and must keep on turning—for ever. It is death if it stands still." [*Brave New World*, p. 48]

Science fiction and social realism confronted the machine directly but literally. As the Century of Progress unfolded, however, even the most visionary flights were quickly overtaken by developments in everyday life. Literature was pressed to become as experimental as engineering, and writers began to deal not just with the objective facts of technology, but with the subjective landscape that had been transformed by it.

The spectral characters of Franz Kafka, who shift through an automatic universe of lifelike shadows and surreal nightmares, probe the soul of the technological era. Kafka, writing at the same historical moment as Max Weber, created a fully routinized world, but one whose administrative machinery is subject to frequent breakdowns. The bureaucratic apparatus of *The Trial* has no point of reference outside of itself, and consumes everything in its path. The centerpiece of *The Penal Colony* is a remarkable piece of judicial machinery, bristling with cogwheels and needles, that executes sentences—and prisoners—with brutal precision; inevitably, both the warden and his machine are destroyed in a single mechanical convulsion.

Bureaucratically integrated and hierarchically controlled, the whole of Kafka's world is one vast penal colony in which all of us are implicated and "guilt is never to be doubted." [*Penal Colony*, p. 145] The sentences of the condemned are literally inscribed on their bodies, as if to affirm that our senses, our perceptions, and our innermost drives have been scripted by the machine.

The same theme recurs a few years later in the novels of D.H. Lawrence, himself the son of a miner, who observed how "the iron and the coal had eaten deep into the bodies and souls of the men." [*Lady Chatterley's Lover*, p. 149] But where Kafka studied the mechanization of the soul, Lawrence grappled with the industrialization of love and sex. Sir Clifford Chatterley is shipped home from Flanders in the last year of the war, "more or less in bits." Paralyzed from the waist down and imprisoned in a motorized wheelchair, he retires to the Midlands, where his "manliness" is restored by the task of reviving the inefficient colliery on his ancestral estate. While her husband is consumed by "this other weirdness of industrial activity," young Constance Chatterley is driven into the arms of their groundskeeper.

The groundskeeper Mellors, "Lady Chatterley's lover," is the only character in the novel who stands outside the industrial system, and the only one capable of arousing in her an erotic passion. Ultimately even he can find no real fulfillment: "It was not the woman's fault," he reflects, "nor even love's fault, nor even the fault of sex. The fault lay there, out there, in those evil electric lights and diabolical rattlings of engines. . . . There lay the vast evil thing, ready to destoy whatever did not conform." [p. 111]

Lawrence never found a satisfactory accommodation with modernity, but lost himself instead in a religio-mystical quest for what he called "the resurrection of the body." He advanced the insight, however, into the means by which the technological apparatus seizes hold of the body, ravages it, and then casts it aside.

By the time of Lawrence and Kafka, the marriage of science and technology had produced its first industrial offspring—the infant electrical and chemical industries; the engineering curriculum had been formalized, and the professional engineer had become an accepted, even an honored figure in modern life and letters. It was almost predictable that two trained engineers, Robert Musil in Austria and Evgeny Zamiatin in the Soviet Union, would come forward with their precision-machined literary masterpieces. Without sacrificing the critical function of literature, the technological imagination helped to dispose of the idea that machines and structures are enemies of art. To the contrary, the engineered environment could provide the writer with metaphors, images, and even models. William Carlos Williams understood this when he wrote that "a poem is a small (or a large) machine made of words." [*The Wedge*, 1944]

Indeed, the aesthetic dimension of the manufactured, built, and designed environment had been a growing preoccupation for artists ever since William Morris conjured in his fertile imagination a civilization built upon aesthetic foundations. Morris, it will be recalled, regarded the machine as the greatest obstacle to the integration of work, justice, and beauty. In the hands of his 20th-century successors, however, technology became a vital source of artistic expression.

The new industrial technologies heralded both a crisis and an opportunity for art, which were felt by the end of the 19th century. In 1888 the civil engineer Gustave Eiffel began construction of a graceful, thousand-foot tower to mark the centenary

Figure 6. 1

A poem is a small (or a large) machine made of words.

William Carlos Williams

Details of the arch span, tracing, Gustave Eiffel

of the French Revolution. Structurally the purpose of the Eiffel Tower was to hold itself up; but symbolically it served as a confident assertion of the new art of structural engineering. The gradual perfection of photography over the course of several decades raised an even more powerful challenge to received notions of art: What did it mean if an untrained amateur, armed with a mechanical contraption, could in seconds create a more "accurate" image than the most skilled academic painter? In 1908 the photographer Alfred Stieglitz opened *Gallery 291* on Fifth Avenue and announced to the world that yet another new art form had been born, and that it was the stepchild of technology.

But modern technology did more than provide new media with which the visual artist, the composer, the architect and designer could experiment. It also helped to free the artistic imagination from the inherited canons of art itself. It is no accident that the invention of photography was followed closely by the birth of Impressionism and the loosening up of the representational ethos to which painting had been bound since the Renaissance. In this respect the great movements of the 20th-century *avant-garde*—Cubism, Expressionism, Surrealism—which display so proudly the powers of the unframed imagination, can be seen as indirect responses to the perfection of the camera.

The Russian Constructivists and the Italian futurists responded more directly. Futurism, which had its origins in the visionary ideas of the Italian poet Filippo Marinetti, boldly confronted the changed horizons of 20th-century art in contrast to the lugubrious fatalism that permeated the cultural landscape of Germany. The Futurists seized upon the technological dynamism of modernity and celebrated the new art, the new society, indeed the new human type, that it heralded. The raw power of the machine would free art from the dead hand of the past, its speed would prevent a new technique from ossifying into a new orthodoxy, its massive scale would defy the narrow cloisters of museums, galleries, and libraries: "We affirm that the world's magnificence has been enriched by a new beauty," Marinetti cried in the first Futurist Manifesto. "A racing car

Figure 6.2

Nothing is more beautiful than a great humming central electric station that holds the hydraulic pressure of a mountain chain and the electric power of a vast horizon.

F.T. Marinetti, *Geometric and Mechanical Splendor*, 1914

Antonio Sant' Elia,
Electric Power Station (1914)

whose hood is adorned by great pipes, like serpents of explosive breath—a roaring car that seems to ride on grapeshot is more beautiful than the *Victory of Samothrace.*" [*Manifesto*, p. 21]

The effusions of the Futurists ranged from the sublime to the ridiculous—and beyond. They denounced museums as "cemeteries of empty exertion, Calvaries of crucified dreams"; Umberto Boccioni called for a new sculpture that would symbolize the violent conquest of space, and Antonio Sant' Elia called for an architecture based on "the superb grace of the steel beam." At the heart of the Futurist program was an ecstatic embrace of mechanical power:

> We will sing of factories hung on clouds by the crooked lines of their smoke; bridges that stride the rivers like giant gymnasts, flashing in the sun with a glitter of knives; adventurous steamers that sniff the horizon; deep-chested locomotives whose wheels paw the tracks like the hooves of enormous steel horses . . ."

[*First Futurist Manifesto*, p. 22]

The *avant-garde* provocations of the Futurists ended, fittingly, in the smoke and fire of World War I, which cost the movement some of its most creative talent. Marinetti, who had written that war is "the world's only hygiene," survived long enough to throw his support behind the Italian Fascist movement and its promise of mobilizing the whole of society's technological resources in a vast, mechanized campaign: "Poets and artists of Futurism!" he cried in a voice now tinged with insanity: "War is beautiful because it initiates the dreamt-of metallization of the human body. War is beautiful because it enriches a flowering meadow with the fiery orchids of machine guns. War is beautiful because it creates new architecture, like that of the big tanks . . . "

A new architecture had indeed come into being by the time of Marinetti's hysterical refrain, but it was grounded in a more rational, humane, and ultimately more successful attempt to realize the aesthetic possibilities of the machine age. Its

Figure 6.3

We must create the mass-production spirit.

Le Corbusier, 1923

Canadian Grain Elevator,
from *Towards a New Architecture*

Barry M. Katz
■ ■ ■ ■ ■ ■

manifesto, at times no less intemperate, was written by the Swiss-French architect Le Corbusier, who in 1923 demanded that the engineer be enlisted in the campaign to free architecture from its humiliating enslavement to past styles. When our machines become obsolete, he reasoned, they are discarded without a second thought and replaced. Why, then, do we continue to erect our buildings in the image of Imperial Rome or Medieval France or Elizabethan England?

Le Corbusier's answer, blunt, disarming, and unequivocal, is that we have not recognized that "the house is a machine for living in," no less than an airplane is a machine for flying in and an automobile a machine for driving in. Just this insight slumbers in the machines and structures of modern engineering. Like William Morris, Le Corbusier believed that every object in our environment is worthy of the artist's hand; but whereas Morris saw the system of mass-production as the greatest threat to both art and industry, Corbusier regarded machine technology as the instrument of liberation: *"Il faut créer l'etat d'esprit de la série!"* he demanded; "We must create the mass-production spirit." [*Toward a New Architecture*, p. 4, 6]

The year 1923 figures prominently in the rapprochement between art and engineering that would prove central to the 20th century. In Paris Le Corbusier published his uncompromising "Purist" manifesto, which called upon the engineer to redirect the course of modern architecture. And in August of that year, in the old German town of Weimar, the architect Walter Gropius was finally prodded into mounting the first public exhibition of the works of the experimental school of design he had founded four years earlier. Under the title "Art and Technology: A New Unity," the *Bauhaus* proposed a revolutionary synthesis that promised to fulfill the needs of modern society by making use of its resources.

The German Bauhaus was both victim and beneficiary of the protracted crisis that stretched from the Treaty of Versailles to the rise of the Nazis. Its uncompromising modernism made it suspect to the political right, and its insistence that art must make common cause with industry was anathema to artistic conservatives. In 1925 the school was forced out of Weimar to the industrial city of Dessau, where, in a new set of buildings designed by Gropius himself, it established itself as the most influential school of design of the 20th century. In the early 1930s the Bauhaus was forced to relocate once again, this time to an abandoned telephone factory in Berlin; finally, like so many of the other leading cultural institutions of Germany, it was driven by the Nazis into what became a permanent American exile. Just this ordeal, however, compelled the artists and artisans of the Bauhaus to formulate a clear statement of their program.

Gropius understood that the future of art and architecture lay not in the repudiation, but in the creative acceptance of precisely those features of industrialism that had proven so alarming to the artistic sensibility: mechanization, the specialized division of labor, standardization of products and the rationalization of production, the new synthetic materials including steel, concrete, and the "sparkling insubstantiality" of glass. To achieve these goals Gropius and his colleagues devised a curriculum that sought to mediate between the "stern realities" of the workplace and the "misty aestheticism" of the academies; their mission was to "avert mankind's enslavement by the machine by giving its products a content of reality and significance." [*Bauhaus*, p. 54]

An adventurous generation of students from all parts of Europe flocked to the Bauhaus, where they were subjected to theoretical courses on the principles of design, but also to workshops in which they had to learn from direct experience about the nature of materials. They were encouraged to cultivate simplicity, economy, and "a decidedly positive attitude to the living environment of vehicles and machines." Everything from desk lamps to stage sets to typefaces to textiles found its way into the curriculum, all of which were to be unified within the overarching domain of architecture. With the coming of the Nazis, they carried the educational principles of the Bauhaus into exile, where they were transplanted in fertile American soil.

Barry M. Katz
■ ■ ■ ■ ■ ■

Thus the modernist ideas of Gropius, Le Corbusier, and their brethren, rooted in the functionalist aesthetic of the engineer, became the religion of the 20th century—with the usual array of heretics, apostates, martyrs, and fanatics. In the hands of lesser builders, the "International Style" visited upon the urban landscape a sterile monotony of flat roofs and inscrutable glass façades. Its potential for greatness was preserved in the mathematical purity of Mies van der Rohe's structures, in the machine-like precision of Marcel Breuer's tubular chairs, and in the quest for elegance, economy, and efficiency that has been admired by even their most hostile critics.

More fundamentally, the modernist revolt helped to promote the acceptance of machine technology and to make the case that craftsmanship can remain a vital force in a world of mechanized mass-production. This credo was sanctified on March 5, 1934, when the Museum of Modern Art opened its "Machine Art" exhibition: Into the Holy of Holies went springs, ball bearings, and various machine components, unidentifiable out of context but beautiful in form.

Technology's critics have not been silenced by its many creative engagements with the world of art and ideas. Nor should they be. Our need for critical guidance grows in proportion to our technical capability: to despoil as well as restore the environment; to wage war and to avert it; to aggravate as well as reduce social inequalities; to stimulate our sensibilites or deaden them. Since the end of World War II—the first to be fought with airplanes and atomic bombs—the world has been haunted by the destructive potential implicit in Francis Bacon's dictum that "knowledge is power."

Awareness of this potential has stimulated the study of technology as a cultural force and made it one of the most vital fields of contemporary thought. It has been noticed, at last, that technology is a human affair, and results not only from the application of scientific law but from chance, luck, mistake, intuition, and poetic imagination; that technology has a history and is, at least in part, a "social construction" that is both a product of specific circumstances and an influence on them.

Most intriguing, perhaps, are recent efforts to broaden and even to rethink fundamentally the meaning of technology. The French philosopher Michel Foucault proposed that psychiatry, criminology, and all systems of thought that authorize acceptable limits of behavior should be thought of as "technologies of the self." Critics of the social inequities, as opposed to biological differences, between the sexes have introduced the concept of "technologies of gender." Media analysts speak of stereotypical depictions of race, class, and gender as "political technologies." No less than water wheels and microchips, these are products of deliberate engineering. We construct a world, and that world constructs us.

Like the tower builders of Babylon, we continue to project ourselves into the sky—and to the end of the solar system, the interior of the atom, the structure of the human genome. But where the ancient engineers harbored only the modest ambition to stand on an equal footing with God, we have more grandiose designs. "Man has, as it were, *become* a prosthetic God."

The Prosthetic God
■ ■ ■ ■ ■ ■

Conclusion

■ ■ ■ ■ ■ ■

A Historical Romance

Technology and Culture have been

together for a long time. Like all romances,

this one has had its high points and its lows;

but despite chronic marital discord,

frequent bouts of infidelity, and even

occasional attempts at separation, this is a relationship that shows no sign of weakening.

The evidence of our five-thousand-year chronicle lends disturbing credence to the words of the preacher in Ecclesiastes: "The sun riseth, and the sun setteth, and there is nothing new under the sun." Like the Sumerian king Gilgamesh, we are still concocting things mechanical and things biological in the same desperate attempt to fend off the inevitable—and with approximately the same results. We are still building towers with their tops in heaven, and suffering the same confusion of tongues that beset our biblical ancestors. No less than Francis Bacon, we are so impressed with our technological virility that we rarely stop to think about whether we are violating the only planet currently available to us. Having learned nothing from Victor Frankenstein, we still forge blindly ahead and hope we will be able to live with whatever we create.

But in the romance of technology and culture there are variables as well as constants. Like the past itself, technology is always changing. The more we implement our technological dreams, the more they return to confront us with new challenges, threats, and opportunities. We have by no means reached the end of the story, then, as if the riddle of technology had now been solved.

The day is long since past when one had to decide whether to cast one's lot with "pro-technological" or "anti-technological" forces. Serious thinkers understand that technology, for better or worse, is part of the human condition, that it always has been, and that it presumably always will be. The task at hand is to render it serviceable to human life.

Bibliography

■ ■ ■ ■ ■ ■

Chapter 1: Technology and the Origins of Culture

I.   Primary sources cited:

The Ancient Near East, vols. I and II, ed. Pritchard (1958).

Poems of Heaven and Hell from Ancient Mesopotamia, trans. Sandars (1971).

The Epic of Gilgamesh, trans. Kovacs (1989).

The Ancient Egyptians: A Sourcebook of their Writings, ed. Adolf Erman (1966).

Documents from Old Testament Times, ed. Thomas (1958).

The Hebrew Bible (various translations including Fox, 1983; N.J.V., 1962; Soncino, 1962).

II.   Further reading:

J.D. Bernal, *Science in History*, vol. I (Cambridge, 1971).

Joseph Campbell, *The Hero with a Thousand Faces* (Princeton, 1968).

Carlo Cipolla and Derek Birdsall, *The Technology of Man* (N.Y., 1979).

V. Gordon Childe, *Man Makes Himself* (N.Y., 1951).

Mircea Eliade, *The Forge and the Crucible* (N.Y., 1971).

Henri Frankfort *et al.*, *The Intellectual Adventure of Ancient Man* (Chicago, 1946).

Siegfried Giedion, *The Beginnings of Architecture* (Princeton, 1981).

Lewis Mumford, *The Myth of the Machine*, vol. I (N.Y., 1966).

John A. Wilson, *The Culture of Ancient Egypt* (Chicago, 1951).

Chapter 2: The Classical Age

I.   Primary sources cited:

The Iliad of Homer (8th century B.C.?), trans. Lattimore (1951).

The Odyssey of Homer (8th century B.C.?), trans. Lattimore (1965).

Hesiod (8th century B.C.?), *Works and Days*, trans. Wender (1973).

Aeschylus (525–456 B.C.), *Prometheus Bound*, trans. Velacott (1961).

Plato (c. 427–347 B.C.),*The Republic*, (various translations including Jowett, Bloom, Cornford); *Philebus*, trans. Hackforth (1961); and *Gorgias*, trans. Woodhead (1961).

Xenophon (c. 430–354 B.C.), *Oeconomicus*, trans. Lord (n.d.).

Aristotle (384–322 B.C.), *Physics, Metaphysics, Ethics*, and *Politics*, ed. McKeon (1941).

Hippocrates (c. 460–377 B.C.), *Precepts*, trans. Jones (1957).

Herodotus (c. 485–c. 425 B.C.), *The Histories*, trans. de Sélincourt (1954).

Arrian (2nd century A.D.), *The Campaigns of Alexander*, trans. de Sélincourt (1958).

Sappho (6th century B.C.?), trans. Barnard (1958).

Plutarch's Lives (c. 46–c. 125 A.D.), vol. V, trans. Perrin (1917).

Virgil (70–19 B.C.), *The Aeneid*, trans. Fitzgerald (1981).

Pliny the Elder (23–79 A.D.), *Natural History*, vol. I, trans. Rackham (1944).

Vitruvius (1st century B.C./A.D.), *The Ten Books on Architecture*, trans. Morgan (1960).

Suetonius (b. 69 A.D.?), *The Lives of the Twelve Caesars*, trans. Gavorse (1931).

Frontinus (c. 35–c. 103 A.D.), *Strategems and Aqueducts*, trans. Bennett (1980).

II.   Further Reading:

Benjamin Farrington, *Greek Science*, vols. I and II (Harmondsworth, 1949); and *Head and Hand in Ancient Greece* (London, 1947).

Moses Finley, *Economy and Society in Ancient Greece* (London, 1981).

J.G. Landels, *Engineering in the Ancient World* (Berkeley, 1978).

G.E.R. Lloyd, *Early Greek Science* (N.Y., 1970); and *Greek Science after Aristotle* (N.Y., 1973).

Claude Mossé, *The Ancient World at Work* (N.Y., 1969).

Derek de Solla Price, *Science Since Babylon* (New Haven, 1975).

L. Sprague de Camp, *The Ancient Engineers* (N.Y., 1975).

K.D. White, *Greek and Roman Technology* (Ithaca, 1975).

Chapter 3: Technology and World Culture

I.   Primary sources cited:

The Four Gospels and the Revelation, trans. Lattimore (1979).

St. Augustine (354–430), *The City of God*, trans. Walsh *et al.* (1958).

The Rule of St. Benedict (480–567), trans. Meisel (1975).

The Didascalicon of Hugh of St. Victor (1096–1141), trans. Taylor (1961).

Abbot Suger (1081–1151), *On the Consecration of the Church of Saint-Denis,* trans. Panofsky (1946).

Theophilus (c. 1125), *On Divers Arts*, trans. Smith (1979).

The Sketchbook of Villard de Honnecourt (c. 1235), ed. Bowie (1959).

Roger Bacon (1214–94), *Opus Majus*, trans. Burke (1928); and "Letter Concerning the Marvelous Power of Art and of Nature and Concerning the Nullity of Magic," trans. Davis (1923).

Maimonides (1135–1204), "Letter on Astrology," trans. Lerner, in *Medieval Political Philosophy;* and *A Sourcebook*, ed. Lerner & Mahdi (1963).

The Song of Roland (early 12th century), trans. Terry (1965).

Banu Musâ (9th century), *Kitab al-Hiyal (The Book of Ingenious Mechanical Devices)*, trans. Hill (1979).

Al-'Amiri (d. 991), quoted in Hassan & Hill, *Islamic Technology* (1986).

Ibn al-Razzaz al-Jazârî (12th century), *The Book of Knowledge of Ingenious Mechanical Devices*, trans. Hill (1974).

The Book of Ser Marco Polo (1254–1324), trans. Yule (1871).

Chuang Tzu (3rd century B.C.), quoted in Needham, *The Shorter Science and Civilization in China*, vol. I (1978).

Ibn Khaldûn (1332–1406), *The Muqaddimah: An Introduction to History*, trans. Rosenthal (1967).

Al-Ghazâlî (1058–1111), *Deliverance from Error*, trans. Watt (1953).

Averroës (1126–98), *On the Harmony of Religion and Philosophy*, trans. Hourani (1961).

Christine de Pizan (1365–1430), *The Book of the City of Ladies*, trans. Richards (1982).

II.   Further reading:

A.C. Crombie, *Medieval and Early Modern Science*, vol. I (N.Y., 1959).

Jean Gimpel, *The Medieval Machine* (Harmondsworth, 1976).

Ahmad al-Hassan & Donald Hill, *Islamic Technology* (Cambridge, 1986).

Lewis Mumford, *Technics and Civilization* (N.Y., 1934).

Seyyed Hossein Nasr, *Science and Civilization in Islam* (N.Y., 1968).

Joseph Needham, *The Grand Titration* (Toronto, 1969).

Arnold Pacey, *The Maze of Ingenuity* (N.Y., 1975).

Lynn White Jr., *Medieval Religion and Technology* (Berkeley, 1978).

Chapter 4: Head and Hand in the Culture of the Renaissance

I.   Primary Sources Cited:

Leon Battista Alberti (1404–72), *Ten Books on Architecture*, trans. Leoni (1955).

Niccoló Machiavelli (1469–1527), *The Prince*, trans. Bull (1961).

Leonardo da Vinci (1452–1519), *Notebooks*, vols. I and II, trans. Richter (1970).

Giorgio Vasari (1511–74), *Lives of the Most Eminent Painters, Sculptors, and Architects*, ed. Burroughs (1946).

Vanoccio Biringuccio (1480–1539), *Pirotechnica*, trans. Smith (1990).

Georgius Agricola (1494–1555), *De Re Metallica*, trans. Hoover & Hoover (1950).

Andreas Vesalius (1514–64), *De Humani Corporis Fabrica*, trans. Singer (1952).

Agostino Ramelli (1527–1608?), *The Various and Ingenious Machines*, trans. Gnudi (1976).

Giovanni Pico della Mirandola (1463–94), *Oration on the Dignity of Man*, trans. Caponigri (1956).

Paracelsus (1493–1541), *Selected Writings*, trans. Guterman (1951).

Christopher Marlowe (1564–93), *The Tragicall History of the Life and Death of Doctor Favstvs*, ed. Jump (1962).

Ben Jonson (1572–1637), *The Alchemist*, ed. Kernan (1974).

William Shakespeare (1564–1616), *The Tempest*, ed. Orgel (1987).

Francis Bacon (1561–1626), *Novum Organum*, ed. Anderson (1960); *New Atlantis*, ed. Jones (1937); and *Preparative for a Natural and Experimental History*, ed. Anderson (1960).

René Descartes (1596–1650), *Discourse on Method*, trans. Haldane and Ross (1955).

Jonathan Swift (1667–1745), *Gulliver's Travels*, ed. Landa (1960).

Jean Le Rond d'Alembert (1717–83), *Preliminary Discourse*, trans. Schwab (1963).

Denis Diderot (1713–84) *et al.*, *Pictorial Encyclopedia of Trades and Industry*, ed. Gillespie (1959).

Diderot, d'Alembert, and a Society of Men of Letters, *Encyclopedia*, trans. Hoyt & Cassirer (1965).

Julien Offray de la Mettrie (1709–51), *Man a Machine*, ed. Bussey (1912).

Jean-Jacques Rousseau (1712–78), *Confessions*, trans. Grant (1931); and *Discourse on the Arts and Sciences*, trans. Masters (1964).

II.   Further reading:

E.A. Burtt, *The Metaphysical Foundations of Modern Science* (N.Y., 1954).

Herbert Butterfield, *The Origins of Modern Science* (N.Y., 1957).

Bertrand Gille, *Engineering in the Renaissance* (Cambridge, 1966).

Marjorie Nicolson, *Science and Imagination* (Ithaca, 1956).

Paolo Rossi, *Philosophy, Technology, and the Arts in the Early Modern Era* (N.Y., 1970).

Londa Schiebinger, *The Mind Has No Sex? Women in the Origins of Modern Science* (Cambridge, 1989).

Wayne Shumaker, *The Occult Sciences in the Renaissance* (Berkeley, 1972).

Chapter 5: Technology and Revolution

I.   Primary sources cited:

Johann Wolfgang von Goethe (1749–1832), *Faust*, I and II, trans. Arndt (1976); trans. Passage (1965).

Thomas Carlyle (1795–1881), "Signs of the Times," *Works*, vol. II; and *Past and Present*, ed. Altick (1977).

Percy Bysshe Shelley (1792–1822), *A Defense of Poetry*, ed. Jordan (1965); and *Prometheus Unbound* (n.d.).

Mary Shelley (1797–1851), *Frankenstein, or The Modern Prometheus*, ed. Joseph (1969).

Adam Smith (1723–90), *The Wealth of Nations*, ed. Cannan (1937).

William Blake (1757–1827), *Jerusalem*, in *Complete Writings* (1966).

Friedrich Engels (1820–95), *The Condition of the Working Class in England*, trans. Henderson and Chaloner (1968).

John Stuart Mill (1806–73), *Principles of Political Economy*, ed. Kelly (1965); *The Spirit of the Age* (ed. Schneewind, 1965); and *Autobiography* (1924).

Henri de Saint Simon (1760–1825), *Social Organization, The Science of Man, and other Writings*, trans. Markham (1964).

Charles Fourier (1772–1837), *Selected Texts on Work, Love, and Passionate Attraction,* trans. Beecher and Bienvenu (1971).

Karl Marx (1818–83) and Friedrich Engels (1820–95), *Manifesto of the Communist Party,* ed. Fernbach (1974); and *The German Ideology,* vol. I, ed. Arthur (1970).

Karl Marx, *Capital,* vol. 1, trans. Fowkes (1977).

William Morris (1834–96), "The Lesser Arts," "How We Live and How We Might Live," "Useful Work versus Useless Toil," "A Factory as it Might Be," and "How I Became a Socialist," in *Political Writings,* ed. Cole (1948).

Auguste Comte (1798–1857), *Introduction to Positive Philosophy,* trans. Ferré (1970).

Charles Darwin (1809–82), *The Origin of Species* and *The Descent of Man,* ed. Appleman (1979).

T.H. Huxley (1825–95), *Evolutionary Ethics,* in Applemean (1979).

Friedrich Nietzsche (1844–1900), *The Gay Science,* trans. Kaufmann (1974); *Thus Spoke Zarathustra,* trans. Hollingdale (1969); *Beyond Good and Evil,* trans. Kaufmann (1966); *On the Genealogy of Morals,* trans. Kaufmann (1969); *Twilight of the Idols,* trans. Hollingdale (1968); and *The Will to Power,* trans. Kaufmann and Hollingdale (1968).

II.   Further reading:

Siegfried Giedion, *Mechanization Takes Command* (N.Y., 1969).

Robert Heilbroner, *The Worldly Philosophers* (N.Y., 1953).

Erich Heller, *The Artist's Journey Into the Interior* (N.Y., 1965).

E.J. Hobsbawm, *Industry and Empire* (Harmondsworth, 1969).

Humphrey Jennings, *Pandaemonium: The Coming of the Machine* (N.Y., 1985).

John Kasson, *Civilizing the Machine: Technology and Republican Virtues in America* (Harmondsworth, 1976).

Elting Morison, *From Know-How to Nowhere: The Development of American Technology* (N.Y., 1974).

Raymond Williams, *Culture and Society, 1780–1950* (N.Y., 1958).

Chapter 6: The Prosthetic God

I.   **Primary sources cited:**

Sigmund Freud (1856–1939), *The Interpretation of Dreams*, ed. Brill (1938); *Civilization and Its Discontents*, trans. Strachey (1962); with Josef Breuer, *Studies on Hysteria*, trans. Strachey (1955).

Oswald Spengler (1880–1936), *The Decline of the West*, vol. I: *Form and Actuality*, and vol. II: *Perspectives of World History;* trans. Atkinson (1983).

Max Weber (1864–1920), *The Protestant Ethic and the Spirit of Capitalism*, trans. Parsons (1958); "Science as a Vocation," and "Politics as a Vocation," in *From Max Weber*, trans. Gerth and Mills (1970).

Max Horkheimer (1895–1973), *Eclipse of Reason* (1974); with Theodor W. Adorno (1903–69), *Dialectic of Enlightenment*, trans. Cumming (1972).

Herbert Marcuse (1898–1979), *Negations: Essays in Critical Theory*, trans. Shapiro (1968); and *One-Dimensional Man* (1964).

Erich Fromm (1900–80), *The Crisis of Psychoanalysis* (1970); and *The Anatomy of Human Destructiveness* (1973).

Walter Benjamin (1892–1940), *Illuminations*, ed. Arendt (1969).

Edmund Husserl (1859–1938), *Phenomenology and the Crisis of Philosophy*, trans. Lauer (1965); and *The Crisis of the European Sciences*, trans. Carr (1970).

Martin Heidegger (1889–1976), *Being and Time*, trans. Maquarrie and Robinson (1962); *What is Called Thinking?*, trans. Wieck and Gray (1972); "The Question Concerning Technology" and "The Age of the World Picture" in *The Question Concerning Technology and Other Essays*, trans. Lovitt (1977); and ". . . Poetically Man Dwells . . ." in *Poetry, Language, Thought*, trans. Hofstadter (1971).

William Carlos Williams (1883–1963), *The Wedge* (1944); and *Collected Poems* (1986).

Jules Verne (1828–1905), *From the Earth to the Moon* (1978); and *Voyage to the Bottom of the Sea* (1943).

H.G. Wells (1866–1946), *The Time Machine* (1987); and *War of the Worlds* (1961).

Emile Zola (1840–1902), *Germinal*, trans. Tancock (1954).

Upton Sinclair (1878–1968), *The Jungle* (1980).

Henry Adams (1838–1918), *The Education of Henry Adams: An Autobiography* (1918).

Franz Kafka (1883–1924), *The Complete Stories*, ed. Glatzner (1946).

D.H. Lawrence (1885–1930), *Lady Chatterley's Lover* (1959).

Evgeny Zamiatin (1884–1937), *We*, trans. Zilboorg (1952).

Robert Musil (1880–1942), *The Man Without Qualities*, trans. Wilkins and Kaiser (1965).

Aldous Huxley (1894–1963), *Brave New World* (1964).

Filippo Tomasso Marinetti (1876–1944) *et al.*, *Futurist Manifestos*, ed. Apollonio (1973).

Le Corbusier (1887–1965), *Towards a New Architecture*, trans. Etchells (1986).

Walter Gropius (1883–1969), *The New Architecture and the Bauhaus*, trans. Shand (1965); *Scope of Total Architecture*, (1943); with Herbert Bayer and Ise Gropius, *Bauhaus, 1919–1928* (1975).

Michel Foucault (1926–84), *Technologies of the Self* (1988); and *The History of Sexuality*, vol. I, trans. Hurley (1980).

II. Further Reading

Reyner Banham, *Theory and Design in the First Machine Age* (Cambridge, 1980).

J. David Bolter, *Turing's Man: Western Culture in the Computer Age* (Chapel Hill, 1984).

David P. Billington, *The Tower and the Bridge: The New Art of Structural Engineering* (N.Y., 1983).

O. B. Hardison Jr., *Disappearing through the Skylight: Culture and Technology in the Twentieth Century* (N.Y., 1989).

Jeffrey Herf, *Reactionary Modernism: Technology, Culture, and Politics in Weimar and the Third Reich* (Cambridge, 1984).

H. Stuart Hughes, *Consciousness and Society: The Reorientation of European Social Thought, 1890-1930* (N.Y., 1958).

Robert Hughes, *The Shock of the New* (N.Y., 1981).

Thomas P. Hughes, *American Genesis: A Century of Invention and Technological Enthusiasm* (N.Y., 1989).

154

Stephen Kern, *The Culture of Space and Time, 1880–1918* (Cambridge, 1983).

Leo Marx, *The Machine in the Garden: Technology and the Pastoral Idea in America* (N.Y., 1964).

Robert M. Pirsig, *Zen and the Art of Motorcycle Maintenance* (N.Y., 1974).

Cecilia Tichi, *Shifting Gears: Technology, Literature, Culture in Modernist America* (Chapel Hill, 1987).

Langdon Winner, *Autonomous Technology: Technics-out-of-Control as a Theme in Political Thought* (Cambridge, 1977).

Credits

■ ■ ■ ■ ■ ■

Barry Katz was born and raised in Chicago. After receiving his B.A. at McGill University in Montreal and his M.Sc. at London School of Economics, he received his Ph.D. from the History of Consciousness program at the University of California at Santa Cruz. In 1980 he joined the faculty of Stanford University, where he is Senior Lecturer in Stanford's Program in Values, Technology, Science and Society and in the Design Division of the Department of Mechanical Engineering.

At Stanford Dr. Katz has taught courses on the aesthetic dimension of modern technology, on technology and warfare, and on theories of technology and culture. A member of the task force that designed Stanford's new Cultures, Ideas, and Values program, he directs a sequence "Technology and Culture," which surveys the cultural significance of technology from stone axes to recombinant DNA. He somehow found time to be co-dean of the Stanford Alumni College in 1989, a responsibility he has agreed to take on again in 1991.

Dr. Katz's research interests are remarkably wide-ranging. His first book, an intellectual biography of the social philosopher Herbert Marcuse, was based on his doctoral dissertation. A second book, a study of academic scholars who worked in the Office of Strategic Services during World War II, was published in 1989 by the Harvard University Press. He is at work on a history of robots.

He rides a very large black motorcycle.

The Portable Stanford Book Series

This is a volume of the Portable Stanford Book Series, published by the Stanford Alumni Association. Subscribers receive each new Portable Stanford volume on approval. The following books may also be ordered, by number, on the adjoining card:

$10.95 titles

- *Technology and Culture: A Historical Romance* by Barry Katz (#4057)
- *2020 Visions: Long View of a Changing World* by Richard Carlson and Bruce Goldman (#4055)
- *"What Is to Be Done?" Soviets at the Edge* by John G. Gurley (#4056)
- *Notable or Notorious? A Gallery of Parisians* by Gordon Wright (#4052)
- *This Boy's Life* by Tobias Wolff (#4050)
- *Ride the Tiger to the Mountain: T'ai Chi for Health* by Martin and Emily Lee and JoAn Johnstone (#4047)
- *Alpha and Omega: Ethics at the Frontiers of Life and Death* by Ernlé W.D. Young (#4046)
- *Conceptual Blockbusting* (3rd edition) by James L. Adams (#4007)

$9.95 titles

- *In My Father's House: Tales of an Unconformable Man* by Nancy Huddleston Packer (#4040)
- *The Imperfect Art: Reflections on Jazz and Modern Culture* by Ted Gioia (#4048)
- *Yangtze: Nature, History, and the River* by Lyman P. Van Slyke (#4043)
- *The Eagle and the Rising Sun: America and Japan in the Twentieth Century* by John K. Emmerson and Harrison M. Holland (#4044)
- *The Care and Feeding of Ideas* by James L. Adams (#4042)
- *The American Way of Life Need Not Be Hazardous to Your Health* (Revised Edition) by John W. Farquhar, M.D. (#4018)
- *Cory Aquino and the People of the Philippines* by Claude A. Buss (#4041)
- *50: Midlife in Perspective* by Herant Katchadourian, M.D. (#4038)
- *Under the Gun: Nuclear Weapons and the Superpowers* by Coit D. Blacker (#4039)
- *Wide Awake at 3:00 A.M.: By Choice or By Chance?* by Richard M. Coleman (#4036)
- *Hormones: The Messengers of Life* by Lawrence Crapo, M.D. (#4035)
- *Panic: Facing Fears, Phobias, and Anxiety* by Stewart Agras, M.D. (#4034)

- *Who Controls Our Schools? American Values in Conflict* by Michael W. Kirst (#4033)
- *Matters of Life and Death: Risks vs. Benefits of Medical Care* by Eugene D. Robin, M.D. (#4032)

$8.95 titles
- *Terra Non Firma: Understanding and Preparing for Earthquakes* by James M. Gere and Haresh C. Shah (#4030)
- *On Nineteen Eighty-Four* edited by Peter Stansky (#4031)
- *The Musical Experience: Sound, Movement, and Arrival* by Leonard G. Ratner (#4029)
- *Challenges to Communism* by John G. Gurley (#4028)
- *Cosmic Horizons: Understanding the Universe* by Robert V. Wagoner and Donald W. Goldsmith (#4027)
- *Beyond the Turning Point: The U.S. Economy in the 1980s* by Ezra Solomon (#4026)
- *The Age of Television* by Martin Esslin (#4025)
- *Insiders and Outliers: A Procession of Frenchmen* by Gordon Wright (#4024)
- *Mirror and Mirage: Fiction by Nineteen* by Albert J. Guerard (#4023)
- *The Touch of Time: Myth, Memory, and the Self* by Albert J. Guerard (#4022)
- *The Politics of Contraception* by Carl Djerassi (#4020)
- *Economic Policy Beyond the Headlines* by George P. Shultz and Kenneth W. Dam (#4017)
- *Tales of an Old Ocean* by Tjeerd van Andel (#4016)
- *Law Without Lawyers: A Comparative View of Law in China and the United States* by Victor H. Li (#4015)
- *The World That Could Be* by Robert C. North (#4014)
- *America: The View from Europe* by J. Martin Evans (#4013)
- *An Incomplete Guide to the Future* by Willis W. Harman (#4012)
- *Murder and Madness* by Donald T. Lunde, M.D. (#4010)
- *The Anxious Economy* by Ezra Solomon (#4009)
- *The Galactic Club: Intelligent Life in Outer Space* by Ronald Bracewell (#4008)
- *Is Man Incomprehensible to Man?* by Philip H. Rhinelander (#4005)
- *Some Must Watch While Some Must Sleep* by William E. Dement, M.D. (#4003)
- *Human Sexuality: Sense and Nonsense* by Herant Katchadourian, M.D. (#4002)

☐ YES, **please send me ___copy(ies) of** _____
_____ (or books checked on the enclosed list) at $8.95/
9.95/$10.95 each. California residents add 7¼ percent tax per book.
Please add $2.25 per order for shipping and handling.

Name_____

Address_____

City, State, Zip_____

☐ **Please send gift copy(ies) of**_____ **to:**

Gift Recipient Name_____

Address_____

City, State, Zip_____

Gift Recipient Name_____

Address_____

City, State, Zip_____

☐ Payment enclosed.   ☐ Bill my Mastercard/Visa (circle one)

Account no. _____Exp. _____

☐ YES, **I would like to become a Portable Stanford
subscriber** and automatically receive: three to four titles a year,
free postage and handling, the **PS newsletter,** and a **free slipcase**
to hold four Portable Stanford volumes. If I don't wish to keep a
book, I can simply place it back in its mailing carton and Portable
Stanford will pay return postage.

Name_____

Address_____

City, State, Zip_____

☐ Please send me The Portable Stanford Catalogue.

☐ Please send me information on The Portable Stanford
Gift Subscription offer.

Do your friends love to read? We will send them information
on The Portable Stanford Book Club with your compliments.
Just fill in their names on the other side of this postage-paid
card.

☐ **Yes, please send information on The Portable Stanford to the following people:**

Please Print

Name _____

Address _____

City, State, Zip _____

Name _____

Address _____

City, State, Zip _____

Name _____

Address _____

City, State, Zip _____

Your name _____

Fold here and staple or tape

The Portable Stanford
Stanford Alumni Association
Bowman Alumni House
Stanford, CA 94305